WORKSHEETS
FOR CLASSROOM OR LAB PRACTICE

CARRIE GREEN

INTERMEDIATE ALGEBRA THROUGH APPLICATIONS

SECOND EDITION

Geoffrey Akst
Borough of Manhattan Community College, City University of New York

Sadie Bragg
Borough of Manhattan Community College, City University of New York

PEARSON
Addison
Wesley

Boston San Francisco New York
London Toronto Sydney Tokyo Singapore Madrid
Mexico City Munich Paris Cape Town Hong Kong Montreal

Reproduced by Pearson Addison-Wesley from electronic files supplied by the author.

Copyright © 2009 Pearson Education, Inc.
Publishing as Pearson Addison-Wesley, 75 Arlington Street, Boston, MA 02116.

ISBN-13: 978-0-321-53641-9
ISBN-10: 0-321-53641-X

1 2 3 4 5 6 OPM 11 10 09 08

Table of Contents

Chapter 1   ALGEBRA BASICS

1.1 Introduction to Real Numbers

Learning Objectives
1   Identify whole numbers, integers, rational numbers, irrational numbers, and real numbers.
2   Graph real numbers on the number line.
3   Find the additive inverse and the absolute value of real numbers.
4   Compare real numbers.
5   Solve applied problems involving real numbers.

Key Terms

Use the most appropriate term or phrase from the given list to complete each statement in exercises 1-5.

real numbers	equation	rational numbers	integers
absolute value	infinity	additive inverses	multiplicative inverses
inequality	sign	whole numbers	irrational numbers

1. The statement $4 \neq 7$ is an example of a(n) _____.

2. -12 and $\sqrt{11}$ are both _____.

3. _____ are real numbers that have decimal representations that either terminate or repeat.

4. The _____ of a number is its distance on the number line from 0.

5. The _____ are all of the whole numbers and their opposites.

Objective 1   Identify whole numbers, integers, rational numbers, irrational numbers, and real numbers.

Use the following list of numbers for exercises 1–5.

$$-14,\ 0,\ 80,\ 8.74,\ \sqrt{13},\ 8\frac{1}{2},\ -\frac{18}{19}$$

1. Identify the whole numbers.   1._____

2. Identify the integers.   2._____

3. Identify the rational numbers.   3._____

4. Identify the irrational numbers.   4._____

5. Identify the real numbers.   5._____

Objective 2 Graph real numbers on the number line.

Graph on the number line.

6. 2

6.

7. 3.4

7.

8. $-\dfrac{9}{4}$

8.

9. -1.6

9.

10. $\dfrac{5}{2}$

10.

Objective 3 Find the additive inverse and the absolute value of real numbers.

Find the opposite of each number.

11. -63 11._____

12. 57.6   12._____

Find the additive inverse of each number.

13. -26.3

14. $\dfrac{1}{7}$

Find the value.

15. $\left| -\dfrac{12}{7} \right|$

16. $-\left| -4.5 \right|$

Objective 4 Compare real numbers.

Indicate whether each inequality is true or false.

17. $-|18| > |-14|$

18. $-|7| < |-13|$

Fill in each blank with <, >, or = to make a true statement.

19. -7 ___ 7

20. -11 ___ -3

21. 5.608 ___ 1.46

Graph the numbers. Then name them in order from least to greatest.

22. 5, −1, 2.2, and −2.3

22._____

23. −$\frac{4}{5}$, 4$\frac{1}{2}$, 3, and −$\frac{5}{2}$

23._____

Graph on the number line, and express in interval notation.

24. The set of all real numbers greater than or equal to −1

24._____

25. The set of all real numbers between −4 and 3 including −4 and excluding 3

25._____

Objective 5 Solve applied problems involving real numbers.

Solve.

26. The boiling point of liquid argon is −186 degrees Celsius and the boiling of liquid xenon is −108 degrees Celsius. Which liquid has the higher boiling point?

26._____

27. The number of students x in a kindergarten teacher's class this year is 4 fewer than the number y in last year's class. Write an inequality that compares the two class sizes.

27._____

4

Chapter 1   ALGEBRA BASICS

1.2 Operations with Real Numbers

Learning Objectives
1   Add, subtract, multiply, and divide real numbers.
2   Solve applied problems involving computation with real numbers.

Key Terms
Use the most appropriate term or phrase from the given list to complete each statement in exercises 1-5.

reciprocal	**opposite**	**multiplicative inverse**	**order of operations rule**
product	**negative**	**additive inverse**	**difference**
positive	**distributive property**		

1.   To subtract two numbers is to find their _____.

2.   The product of a positive number and a negative number is _____.

3.   The number $\dfrac{1}{n}$ is the multiplicative inverse or _____ of the number n, $n \neq 0$.

4.   According to the _____, always perform operations within grouping symbols first.

5.   Subtracting 3 from 2 is the same thing as adding the _____ of 3 to 2.

Objective 1 Add, subtract, multiply, and divide real numbers.

Add or subtract.

1. $-64 + 28$ 1._____

2. $65 + (-64)$ 2._____

3. $-2+(-96)$

                                                3._____

4. $3-9$

                                                4._____

5. $-6-4$

                                                5._____

6. $-82.3+54.7$

                                                6._____

7. $-4.2-(-1)$

                                                7._____

8. $-\dfrac{9}{5}+\left(-\dfrac{6}{5}\right)$

                                                8._____

9. $\dfrac{5}{12}+\left(-\dfrac{7}{12}\right)$

                                                9._____

10. $-\dfrac{7}{2}-\dfrac{1}{2}$

                                                10._____

Multiply or divide.

11. $(-4)(-5)$

                                                11._____

12. $(8)(-1.1)$ 12._____

13. $-10 \div (-5)$ 13._____

14. $\dfrac{-64}{-8}$ 14._____

15. $-\dfrac{4}{19} \div \left(-\dfrac{7}{19}\right)$ 15._____

Evaluate.

16. 5^7 16._____

17. $(-4)^2$ 17._____

18. -5^4 18._____

Simplify.

19. $8 + 8 \cdot (4 - 3)$ 19._____

20. $6 \div (-3)(-3)$ 20._____

21. $\dfrac{(-2)(7)-(4)(12)}{8-5}$

21._____

22. $-5\left[3\left(5^2-3\right)\div 6+4+(-2)4\right]+7$

22._____

Find the value of each expression.

23. $|3-10|$

23._____

24. $|5-6|-|-1+9|$

24._____

25. $\sqrt{2+5-4}$

25._____

26. $\sqrt{2(6-4+1)}$

26._____

Objective 2 Solve applied problems involving computation with real numbers.

Solve.

27. A submarine at a depth of 1622 ft ascends to a depth of 841 ft. How far did the submarine ascend?

27._____

28. One day the temperature dropped from 6°F to -19°F. How many degrees did the temperature drop?

28._____

29. A mountain with a base 12,719 ft below sea level rises 24,208 feet. What is the elevation above sea level of its peak?

29._____

Chapter 1   ALGEBRA BASICS

1.3 Properties of Real Numbers

Learning Objectives
1   Use the properties of real numbers.
2   Solve applied problems involving the properties of real numbers.

Key Terms
Use the most appropriate term or phrase from the given list to complete each statement in exercises 1-4.

    additive inverse property   **multiplicative property of 0**   **reciprocals**

    multiplicative identity property   **opposites**   **additive identity property**

    distributive property   **associative property**   **commutative property**

1.   The _____ is used to regroup numbers in a sum or product.

2.   The _____ is used to change the order of numbers in a sum or product.

3.   The _____ is used when a number is multiplied by a sum.

4.   By the _____, the sum of a number and its opposite is 0.

Objective 1 Use the properties of real numbers.

Rewrite each expression using the indicated property of real numbers.

1.   Commutative property of addition: $n + 7$ 1._____

2.   Associative property of multiplication: $6(5x)$ 2._____

3.   Distributive property: $5(y + 2)$ 3._____

4.   Distributive property in reverse: $7u + 7v$ 4._____

Indicate the definition, property, or number fact that justifies each statement.

5. $y + 0 = y$ **5.**_____

6. $4 \cdot x = x \cdot 4$ **6.**_____

7. $y + (5 + 7) = y + (7 + 5)$ **7.**_____

Calculate, if possible.

8. $13 + (-73) + (-50) + (-93)$ **8.**_____

9. $-7.6 + (-87.6) + 90.8$ **9.**_____

10. $-2 \cdot (-6) \cdot (-3)$ **10.**_____

11. $\dfrac{0}{-6}$ **11.**_____

Find each of the following.

12. The multiplicative inverse of $\dfrac{3}{4}$

12._____

13. The additive inverse of -5

13._____

Rewrite each expression without the grouping symbols.

14. $z(1+w)$

14._____

15. $(p+q)3r$

15._____

To prove the given statement, justify each step.

16.

$$\left(\frac{1}{3}\right)(6+3x) = x+2$$

$$\left(\frac{1}{3}\right)(6+3x) = \left(\frac{1}{3}\bullet 6\right)+\left(\frac{1}{3}\bullet 3x\right) \quad \text{(a)}$$

$$= \left(\frac{1}{3}\bullet 6\right)+(1)x \quad \text{(b)}$$

$$= 2+x$$

$$= x+2 \quad \text{(c)}$$

16 a._____

b._____

c._____

Objective 2 Solve applied problems involving the properties of real numbers.

Explain each answer in terms of the appropriate property of real numbers or definition.

17. Elise has several guests. She wants to make enough
    dessert to serve all of them, but her recipe will
    serve only $\dfrac{1}{4}$ of them. By what factor must she
    multiply her recipe so that she has enough
    ingredients to make dessert for all guests with no
    ingredients left over?

17._____

18. Jonah bought a pair of pants for $49 and a shirt for
    $32 and he paid for them with a credit card. He
    then made an $81 payment to his credit card. If his
    credit card balance was 0 before he made these
    purchases, will he have a balance after the payment
    is made (assuming all interest charges are
    deferred)?

18._____

Chapter 1   ALGEBRA BASICS

1.4 Laws of Exponents and Scientific Notation

Learning Objectives
1   Rewrite expressions that contain 0 or negative exponents.
2   Rewrite expressions by using the product and quotient rules of exponents.
3   Rewrite expressions by using the power rule or by raising a product or a quotient to a power.
4   Change a number written in scientific notation to standard notation, and vice versa.
5   Solve applied problems involving the laws of exponents and scientific notation.

Key Terms
Use the most appropriate term from the given list to complete each statement in exercises 1-4.

| **base** | **added** | **zero** | **standard notation** | **subtracted** |

| **multiplied** | **divided** | **exponent** | **scientific notation** |

1.   The number 345,000,000 is written in _____.

2.   In the quotient rule of exponents, the exponents are subtracted when powers with like bases are _____.

3.   Any nonzero number raised to the _____ power is equal to 1.

4.   In the product rule of exponents, the exponents are _____.

Objective 1 Rewrite expressions that contain 0 or negative exponents.

Evaluate.

1. $(-43)^0$ 1._____

2. 6^1 2._____

3. $-2y^0$ 3._____

4. p^0 4._____

Express in terms of a base raised to a positive exponent.

5. 10^{-2} 5._____

6. $-x^{-3}$ 6._____

7. $x^5 y^{-2}$ 7._____

8. $8x^{-2}$ 8._____

Objective 2   Rewrite expressions by using the product and quotient rules of exponents.

Rewrite each expression as a base to a power, if possible.

9. $7^6 \cdot 7^9$ 9._____

10. $\dfrac{6^6}{6^3}$ 10._____

11. $a^2 \cdot b^4$ 11._____

12. $\dfrac{z^8}{z^4}$ 12._____

13. $y^4 \cdot y^8 \cdot y$ 13._____

14. $\left(a^2b^5\right)\left(a^3b^2\right)$ 14._____

Objective 3   Rewrite expressions by using the power rule or by raising a product or a quotient to a power.

Simplify.

15. $\left(m^{14}\right)^2$ 15._____

16. $\left(wz\right)^{-5}$ 16._____

17. $\left(\dfrac{a}{b}\right)^8$ 17._____

18. $-\left(4g^3\right)^4$ 18._____

19. $\left(-\dfrac{f^3}{g^6}\right)^5$ 19._____

Objective 4   Change a number written in scientific notation to standard notation, and vice versa.

Express in scientific notation.

20. $870,000,000,000$ 20._____

21. 0.00000000048 21._____

22. 360,000,000,000,000   22._____

Express in standard notation.

23. 4.6×10^7 23._____

24. 9.69×10^{-8} 24._____

25. 1.34×10^{-5} 25._____

Calculate, writing the result in scientific notation.

26. $\left(2 \times 10^4\right)\left(4 \times 10^4\right)$ 26._____

27. $\left(4.4 \times 10^2\right)\left(5.7 \times 10^{-4}\right)$ 27._____

28. $\left(6.5 \times 10^{-7}\right) \div \left(1.3 \times 10^{17}\right)$ 28._____

Objective 5   Solve applied problems involving the laws of exponents and scientific notation.

Solve.

29. In 1998, the sales for the Automation Company   29._____
were $13,000,000. Write this amount in scientific
notation.

30. The approximate number of cells in a sample is   30._____
1.99×10^5. Write this number in standard form.

Chapter 1   ALGEBRA BASICS

1.5 Algebraic Expressions: Translating, Evaluating, and Simplifying

Learning Objectives
1   Translate algebraic expressions to word phrases and word phrases to algebraic expressions.
2   Evaluate algebraic expressions.
3   Simplify algebraic expressions.
4   Solve applied problems involving algebraic expressions.

Key Terms
Use the most appropriate term from the given list to complete each statement in exercises 1-4.

expression	coefficient	combining	constant	equal	terms
canceling	exponent	variables	evaluate	solve	like

1. In the expression $-3x^2$, the _____ applies to the variable only.

2. Addition and subtraction signs separate _____.

3. Like terms have the same _____ with the same exponents.

4. $7x + 3y$ is an example of a(n) _____.

Objective 1   Translate algebraic expressions to word phrases and word phrases to algebraic expressions.

Translate each algebraic expression to words.

1. $3 + 2s$ 1._____

2. $2x - 6y$ 2._____

3. $7(x - y)$ 3._____

4. $\dfrac{p+q}{10}$

4._____

Express each phrase as an algebraic expression.

5. One-fourth of some number n

5._____

6. The difference between a number x and 3

6._____

7. The ratio of c and d

7._____

8. The sum of p and twice q

8._____

9. 7 more than 2 times x

9._____

10. The product of -15 and the difference between d and 6

10._____

Objective 2 Evaluate algebraic expressions.

Evaluate the given expression for each value of the variable.

11. $2x+7$
    a.   0
    b.   1
    c. -1

11a._____

b._____

c._____

12. $-\dfrac{1}{3}x$
    a.   0
    b.   3
    c. -3

12a._____

b._____

c._____

Evaluate the expression.

13. Find the value of the expression $7x - y$ for $x = 4$ and $y = 28$.

13._____

14. Find the value of the expression $\dfrac{-m+18n}{1s+4}$ for

    $m = 12$, $n = -4$, and $s = 0$.

14._____

15. Find the value of the expression $\dfrac{x}{2(3+y)}$ for

    $x = 10$ and $y = -4$.

15._____

Objective 3   Simplify algebraic expressions.

For each algebraic expression, identify the terms and indicate whether they are like or unlike.

16. $5r - 18r^2$ 16._____

17. $10s - 3s$ 17._____

18. $3y + 3z$ 18._____

Combine like terms, if possible.

19. $12t - t$ 19._____

20. $9x^3 + 6x^3 + 5$ 20._____

21. $-6a^2 - 7a$ 21._____

22. $-7u^3v + 5u^3v$ 22._____

Simplify.

23. $-4(t-3)-10$ 23._____

24. $-(c-91)$ 24._____

25. $67m-52-(2+48m)$ 25._____

26. $-3(t-8)-4(t-6)$ 26._____

Objective 4 Solve applied problems involving algebraic expressions.

27. The interior angles of a triangle have measures   27._____
$x+10$, $2x-3$, and $4x+5$. Write and simplify an
expression for the sum of the interior angles of the
triangle.

28. Jason is half the age of his father and one third the   28._____
age of his grandfather. Write and simplify an
expression for the sum of the three ages if Jason's
age is represented by j.

29. The temperature at midnight was 56°F. By noon the   29._____
temperature had risen y degrees. At sundown, the
temperature was 12 degrees less than it was at noon.
Write and simplify an expression for the
temperature at sundown.

30. On Saturday, an amusement park sold $10b$ ice   30._____
cream bars and $8f$ frozen fruit bars. On Sunday, the
amusement park sold $15b$ ice cream bars, $12f$ frozen
fruit bars, and $6c$ chocolate chip cookies. Write and
simplify an expression for the number of treats sold
at the amusement park over the weekend.

Chapter 2   LINEAR EQUATIONS AND INEQUALITIES

2.1 Solving Linear Equations

Learning Objectives
1 Determine whether a given number is a solution of a given equation.
2 Solve linear equations using the addition property.
3 Solve linear equations using the multiplication property.
4 Solve linear equations using both the addition and multiplication properties.
5 Solve linear equations involving combining like terms and/or parentheses.
6 Solve applied problems involving the addition property, multiplication property, combining like terms, and/or parentheses.

Key Terms
Use the most appropriate term or phrase from the given list to complete each statement in exercises 1-4.

linear equations	evaluation	additive inverse property
equivalent equations	inequality	addition property of equality
distributive property	equation	solution

1. _____ are equations that have the same solutions.

2. To solve linear equations, first use the _____ to clear the equation of parentheses, if necessary.

3. According to the _____, if any real number is added to each side of an equation, the result is an equivalent equation.

4. A linear _____ has the form $ax + b = c$, where a, b, and c are real numbers and $a \neq 0$.

Objective 1   Determine whether a given number is a solution of a given equation.

Determine whether the given value is a solution of the equation.

1. $5x + 7 = 30; 5$ 1._____

2. $9x - 30 = -5; 3$ 2._____

3. $2z - 8 = -6z + 56;\ 8$ 3._____

4. $2y - 5 = 6y - 17;\ 3$ 4._____

5. $4(x - 3) = -5x + 7;\ 6$ 5._____

Objective 2 Solve linear equations using the addition property.

Solve and check.

6. $x + 16 = -10$ 6._____

7. $t - 5 = 12$ 7._____

8. $\dfrac{1}{2} + x = \dfrac{5}{2}$ 8._____

9. $m - 14 = -21$ 9._____

10. $x + (-12) = 22$ 10._____

Objective 3 Solve linear equations using the multiplication property.

Solve and check.

11. $\dfrac{x}{4} = 11$ 11._____

12. $-1.8 = 0.3x$

12._____

13. $-7.2 = 0.8n$

13._____

14. $\dfrac{4}{5}x = \dfrac{7}{3}$

14._____

15. $-\dfrac{1}{3} = \dfrac{5}{2}z$

15._____

Objective 4 Solve linear equations using both the addition and multiplication properties.

Solve and check.

16. $7 + 4x = 7$

16._____

17. $\dfrac{1}{8}y + 4 = -16$

17._____

18. $9 - 3a = a - 14$

18._____

19. $6x - 8x + 6 = -4x$

19._____

20. $3 + 11x - 2 = 9x + 28 - 7x$ **20.**_____

Objective 5 Solve linear equations involving combining like terms and/or parentheses.

Solve and check.

21. $3(5x - 2) = 69$ **21.**_____

22. $5(5x - 2) = 90$ **22.**_____

23. $7(5x + 8) = 12 - (x + 4)$ **23.**_____

24. $5(5x + 9) = 13 - (x + 7)$ **24.**_____

25. $5[7 - 5(6 - x)] - 7 = 7[5(7x - 6) + 4] - 50$ **25.**_____

Objective 6   Solve applied problems involving the addition property, multiplication property, combining like terms, and/or parentheses.

Solve.

26. An appliance store decreases the price of a 19-in. television set 23% to a sale price of $488.18. What was the original price?

26._____

27. The sum of the measures of the angles of any triangle is 180 degrees In triangle *ABC*, angles *A* and *B* have the same measure, while the measure of angle *C* is 51 degrees more than each of *A* and *B*. What are the measures of the three angles?

27._____

28. Two friends leave 1.4 mi apart along the same road. If they walk toward each other at rates that differ by $\frac{1}{2}$ mph and meet 24 minutes later, how fast is each person walking?

28._____

Name: Date:
Instructor: Section:

29. The chocolate factory makes a dark chocolate that is   29._____
30% fat and a white chocolate that is 50% fat. How
many kilograms of dark chocolate should be mixed
with 200 kilograms of white chocolate to make a
ripple blend that is 40% fat?

26

Chapter 2   LINEAR EQUATIONS AND INEQUALITIES

2.3 Solving Linear Inequalities

Learning Objectives
1   Determine whether a given number is a solution of a given inequality.
2   Graph the solutions of linear inequalities on a number line.
3   Solve linear inequalities using the addition and multiplication properties of inequalities.
4   Solve applied problems involving linear inequalities.

Key Terms
Use the most appropriate term or phrase from the given list to complete each statement in exercises 1-6.

negative	equation	rational	evaluate	multiplication property
graph	less than	solution	inequality	distributive property
solve	excludes	includes	positive	less than or equal to

1. Multiplying an inequality by a _____ number does not change the direction of the inequality symbol.

2. The interval notation $(-\infty, 7)$ means all real numbers _____ 7.

3. Just as with solving an _____, we sometimes need to use more than one property to solve an inequality.

4. In interval notation, the right parenthesis) indicates that the interval _____ the right endpoint.

5. To clear the parentheses in the inequality $4(5-x)-2 \leq 1$ use the _____.

6. By the _____ of inequalities, for any real numbers a, b, and c, if $a > b$ and $c < 0$, then $ac < bc$.

Objective 1 Determine whether a given number is a solution of a given inequality.

Determine whether the given value is a solution of the inequality.

1. $x - 3 \geq 8$; 6   1._____

2. $x - 2 \leq 7; \; 9$ 2._____

3. $y - 8 > 2y - 3; \; -3$ 3._____

4. $y - 3 < 2y - 4; \; 2$ 4._____

Objective 2 Graph the solutions of linear inequalities on a number line.

Graph the inequality on the number line. Then write the solution using interval notation.

5. $x < 3$ 5._____

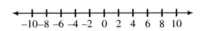

6. $x \geq -5$ 6._____

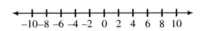

7. $x > 7$ 7._____

8. $x \leq -2$ 8._____

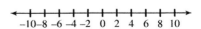

Objective 3   Solve linear inequalities using the addition and multiplication properties of inequalities.

Solve and graph. Then write the solution using interval notation.

9. $x - 7 \geq -1$

9._____

10. $x - 9 \geq -7$

10._____

11. $2x + 1 \geq x + 4$

11._____

12. $-4x < 20$

12._____

13. $-52 \leq -13x$

13._____

14. $-\dfrac{5}{3}x \leq -5$

14._____

15. $-\dfrac{5}{2}x \leq -10$

15._____

Solve.

16. $3 + 5m \leq 73$

16._____

17. $4.3x + 21.6 > 54 - 6.5x$

17._____

18. $3(7 - 6x) + 4x < 5(6 + 3x)$

18._____

19. $5(6 - 3x) + 6x < 8(4 + 2x)$

19._____

20. $\dfrac{1}{2}(6y + 2) - 25 < -\dfrac{1}{4}(8y - 4)$

20._____

Objective 4 Solve applied problems involving linear inequalities.

Solve.

21. A car rents for $30 per day plus $0.18 per mile.
    You are on a daily budget of $75. What mileage
    will allow you to stay within your budget?

22. A student is taking a computer course in which
    there are four tests, each worth 100 points. His
    scores on the first three tests were 89, 93, and 89.
    He must earn a total of 360 points in order to get
    an A. What score on the last test will give him an
    A?

23. Kyle plans to invest $5000, part at 4% simple
    interest and the rest at 5% simple interest. What is
    the most that he can invest at 4% and still be
    guaranteed at least $220 in interest per year?

24. The Warbler House Inn hosts private parties at two different rates. Rate A includes a flat fee of $1700 plus $20 for each guest in excess of the first 40. Rate B simply charges $35 for each guest. For what size parties will using rate B cost less than using rate A?

24._____

Chapter 2   LINEAR EQUATIONS AND INEQUALITIES

2.4 Solving Compound Inequalities

Learning Objectives
1   Solve and graph inequalities containing the word *and*.
2   Solve and graph inequalities containing the word *or*.
3   Solve applied problems involving compound inequalities.

Key Terms
Use the most appropriate term from the given list to complete each statement in exercises 1-4.

intersection   **union**   **expressions**   **inequalities**

equations   **and**   **or**

1.   A solution of a compound inequality joined by the word *and* is the _____ of the solutions of the individual inequalities.

2.   A solution of a compound inequality joined by the word *or* is the _____ of the solutions of the individual inequalities.

3.   A solution to an compound inequality joined by *and* is a solution that makes both _____ true.

4.   The symbol ∪ is used to express the solutions of compound inequalities joined by the word _____

Objective 1 Solve and graph inequalities containing the word *and*.

Solve and graph the inequality. Write the solution using interval notation.

1. $-2x > -12$ and $x + 2 > 0$

    1._____

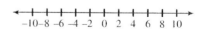

2. $-2x > -14$ and $x + 2 > 0$

    2._____

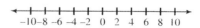

3. $7 < 2y + 11 < 15$

3._____

4. $11 < 3y + 14 \leq 23$

4._____

5. $-2x > -4$ and $x + 6 > 0$

5._____

Solve. Write each solution as an inequality and in interval notation.

6. $-8 < 3 + \dfrac{1}{2}p < 8$

6._____

7. $1 \leq -\dfrac{1}{3}(4x - 27) < 17$

7._____

8. $x - 9 \leq -4 - \dfrac{x}{4}$ and $14x - 17x < -18$

8._____

9. $x - 7 \le -2 - \dfrac{x}{4}$ and $13x - 16x < -21$

9._____

10. $1 \le -\dfrac{1}{2}(3x - 14) < 13$

10._____

Objective 2   Solve and graph inequalities containing the word *or*.

Solve and graph the inequality. Write each solution as an inequality and using interval notation.

11. $4x + 6 \ge -10$ or $6x + 3 \ge -15$

11._____

12. $2x + 4 \ge -14$ or $4x + 2 \ge -6$

12._____

13. $x + 2 > 6$ or $8 - x > 10$

13._____

14. $-19 + 3x < 1 - x$ or $4x - 27 < x$

14._____

15. $x + 7 > 16$ or $2 - x > -3$

15. _____

Solve. Write each solution as an inequality and using interval notation.

16. $x + 3 > 9$ or $3 - x > 0$

16. _____

17. $8 + 2x > 8 - x$ or $5x - 24 > x$

17. _____

18. $3x - 1 > -13$ or $5x + 3 \leq 13$

18. _____

19. $\frac{1}{3}(x + 15) < 8$ or $4(x - 5) > 3x - 16$

19. _____

20. $\frac{1}{5}(x + 21) < 6$ or $3(x - 4) > 2x - 7$

20. _____

Objective 3   Solve applied problems involving compound inequalities.

Solve.

21. The body mass index, *I*, can be used to determine an
individual's risk for heart disease. An index less than
25 indicates a low risk. The body mass index is given
by the formula $I = \dfrac{700W}{H^2}$, where *W* = weight, in
pounds, and *H* = height, in inches. Jerome is 67
inches tall. What weights will keep his body mass
index between 23 and 32?

21._____

22. A Fahrenheit-to-Celsius temperature conversion
formula is $C = \dfrac{5}{9}(F - 32)$. A certain metal is liquid
for Celsius temperatures *C* such that
$751° \leq C < 2842°$. Write the equivalent inequality
for Fahrenheit temperatures.

22._____

23. A cellular phone company offers a contract for
which the cost C, in dollars, of t minutes of
telephoning is given by $C = 0.25(t - 600) + 43.95$,
where it is assumed that $t \geq 600$ minutes. What
times will keep costs between $91.20 and $112.70?

23._____

Chapter 2   LINEAR EQUATIONS AND INEQUALITIES

2.5 Solving Absolute Value Equations and Inequalities

Learning Objectives
1   Solve equations involving absolute value.
2   Solve inequalities involving absolute value.
3   Solve applied problems involving absolute value equations and inequalities.

Key Terms

Use the most appropriate term or phrase from the given list to complete each statement in exercises
1-4.

> **two solutions   plus or minus   that number   add and then subtract**
>
> **no solution   one solution   the opposite of that number**

1.   The solutions of the equation $|x| = 7$ are _____ 7.

2.   The equation $|y| = -1$ has _____.

3.   The equation $|t| = 0$ has _____.

4.   A number and _____ are the same distance from 0 on the number line.

Objective 1   Solve equations involving absolute value.

Solve.

1. $|x| = 19$ 1._____

2. $|g| + 4 = 4$ 2._____

3. $|6x| - 5 = 43$ 3._____

4. $|9x| - 5 = 49$ 4._____

5. $|5y + 5| = 25$ 5._____

6. $|5y + 30| = 35$ 6._____

7. $|9x - 49| = -41$ 7._____

8. $-|3x - 14| = 10$ 8._____

9. $\left|b - \dfrac{31}{7}\right| = \left|\dfrac{6b}{7} - 8\right|$ 9._____

10. $|3x + 8| = |x - 11|$ 10._____

Objective 2 Solve inequalities involving absolute value.

Solve and then graph. Write each solution as an inequality and using interval notation.

11. $|x| < 1$ 11._____

12. $|x| < 7$

12._____

$\longleftrightarrow$ +-+-+-+-+-+-+-+-+-+-+ $\longrightarrow$

13. $|x + 10| \le 5$

13._____

$\longleftrightarrow$ +-+-+-+-+-+-+-+-+-+-+ $\longrightarrow$

14. $|x + 12| \le 10$

14._____

$\longleftrightarrow$ +-+-+-+-+-+-+-+-+-+-+ $\longrightarrow$

15. $|x + 17| > 16$

15._____

$\longleftrightarrow$ +-+-+-+-+-+-+-+-+-+-+ $\longrightarrow$

16. $|x + 11| > 7$

16._____

$\longleftrightarrow$ +-+-+-+-+-+-+-+-+-+-+ $\longrightarrow$

17. $|4 - 4x| \le 3$

17._____

$\longleftrightarrow$ +-+-+-+-+-+-+-+-+-+-+ $\longrightarrow$

18. $|12 - 4x| \le 3$

18._____

$\longleftrightarrow$ +-+-+-+-+-+-+-+-+-+-+ $\longrightarrow$

19. $-|12-4x| \geq -9$

19._____

<----+--+--+--+--+--+--+--+--+--+--+--+--->

20. $-|6-4x| > -5$

20._____

<----+--+--+--+--+--+--+--+--+--+--+--+--->

Objective 3 Solve applied problems involving absolute value equations and inequalities.

Solve.

21. Research at a major university has shown that
identical twins generally differ by less than ± 6
pounds in body weight. If Kim weighs 157 pounds,
then in what range is the weight of her identical
sister Kathy?

21._____

22. The recommended daily intake (RDI) of a
nutritional supplement for a certain age group is
1400 mg/day. Actually, supplement needs vary
from person to person. Write an absolute value
inequality to express the RDI plus or minus 100 mg
and solve it.

22._____

Chapter 3   GRAPHS, LINEAR EQUATIONS AND INEQUALITIES, AND FUNCTIONS

3.1 The Rectangular Coordinate System

Learning Objectives
1   Plot points on the coordinate plane.
2   Identify the quadrants of the coordinate plane.
3   Interpret graphs in applied problems.

Key Terms
Use the most appropriate term or phrase from the given list to complete each statement in exercises 1-5.

horizontal	origin	vertical	ordinate	*y*-axis
coordinate axes	second	**Quadrant II**	**Quadrant IV**	first
x-axis	independent	dependent	ordered pair	

1.   The *x*-coordinate in an ordered pair represents a(n) _____ distance.

2.   The vertical number line on the coordinate plane is called the _____.

3.   Points in _____ lie to the left of the *y*-axis and above the *x*-axis.

4.   The _____ coordinate in an ordered pair is the *y*-coordinate.

5.   Any point on the _____ has a *y*-coordinate of 0.

Objective 1   Plot points on the coordinate plane.

Plot the points with the given coordinates.

1. $A(-2,4)$
     $B(4,5)$
     $C(3,-2)$
     $D(-1,1)$
     $E(-4,-2)$

1.

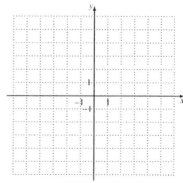

2. $A(-3,-2)$

   $B(2,4)$

   $C(-3,1)$

   $D(-5,-4)$

   $E(1,-4)$

2.

3. $A(3,3)$

   $B(-1,-4)$

   $C(2,5)$

   $D(-4,3)$

   $E(-3,4)$

3.

For each named point, write its coordinates.

4.

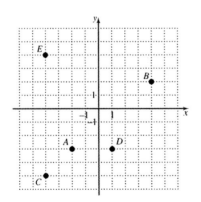

4. _____

   _____

   _____

   _____

   _____

   _____

5.

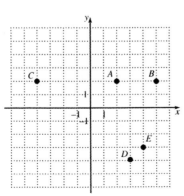

5. _____

   _____

   _____

   _____

   _____

   _____

6.

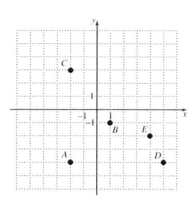

6._____

_____

_____

_____

_____

Objective 2 Identify the quadrants of the coordinate plane.

Identify the quadrant in which each point is located.

7. $(-8,2)$ 7._____

8. $(7.2,6.5)$ 8._____

9. $(-3,-14)$ 9._____

10. $(9,-8.2)$ 10._____

11. $(-9,1)$ 11._____

12. $(5.3,2.7)$ 12._____

13. $(-4,-11)$ 13._____

14. $(7,-9)$ 14._____

15. $(2.1,4.8)$ 15._____

16. $(-6, -12)$ 16._____

Objective 3 Interpret graphs in applied problems.

Solve.

17. The line graph shows the sales (in millions of   17._____
dollars) for several years for a company. What were
the sales for 1995?

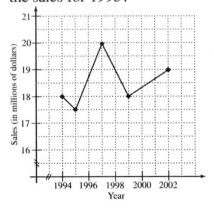

18. The line graph shows estimated sales (in millions of   18a._____
dollars) for a company since 2002.

_____

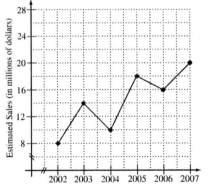
                                                              b._____

_____

a. If (x, y) represents a point on the graph, what
        does x represent?   c._____
b. In what years did the company's estimated sales
        decrease?
c. In what year were estimated sales the greatest?

Chapter 3   GRAPHS, LINEAR EQUATIONS AND INEQUALITIES, AND FUNCTIONS

3.2 Slope

Learning Objectives
1   Find the slope of a line that passes through two given points.
2   Determine whether the slope of a given line is positive, negative, zero, or undefined.
3   Determine whether two given lines are parallel or perpendicular.
4   Interpret slopes in applied problems.

Key Terms
Use the most appropriate term or phrase from the given list to complete each statement in exercises 1-4.

vertical	parallel	decreasing	undefined
increasing	rise	negative	0
horizontal	slant	positive	perpendicular

1.   A line rising from left to right is _____.

2.   A line that is vertical has a slope that is _____.

3.   A line slanting down from left to right has a(n) _____ slope.

4.   A line that is horizontal has a slope that is _____.

Objective 1   Find the slope of a line that passes through two given points.

*Compute the slope **m** of the line that passes through the given points. Then plot the points and sketch the line.*

1. $(0,3)$ and $(-3,-4)$

1._____

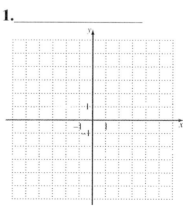

2. $(0,-1)$ and $(2,0)$

2._____

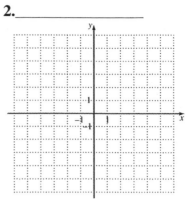

3. $(1,1)$ and $(1,-3)$

3._____

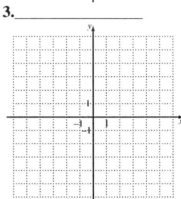

4. $(-1,4)$ and $(4,1)$

4._____

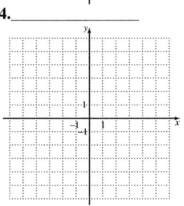

Calculate the slopes of the lines shown.

5.

5._____

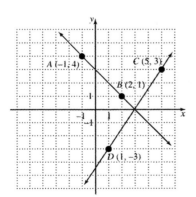

Graph the line on the coordinate plane using the given information.

6. Passes through $(-5,-3)$ and $m = \dfrac{1}{2}$

6.

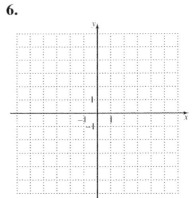

7. Passes through $(1,5)$ and has undefined slope

7.

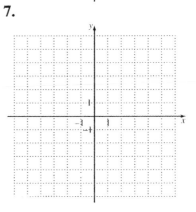

Objective 2 Determine whether the slope of a given line is positive, negative, zero, or undefined.

For each graph identify whether the slope of the line is positive, negative, zero, or undefined.

8.

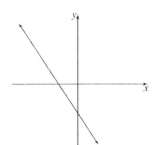

8._____

9.

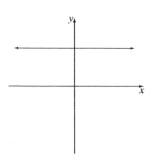

9._____

10.

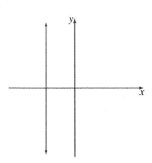

10._____

11.

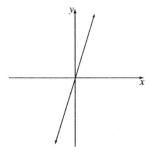

11._____

Objective 3 Determine whether two given lines are parallel or perpendicular.

Given points P, Q, R, and S, determine whether $\overleftrightarrow{PQ}$ and $\overleftrightarrow{RS}$ are parallel, perpendicular, or neither.

12. $P(-1,-5)$, $Q(3,5)$, $R(4,-3)$, and $S(8,7)$

12._____

13. $P(8,3)$, $Q(9,2)$, $R(-6,5)$, and $S(-5,4)$ **13.**_____

14. $P(8,5)$, $Q(10,-10)$, $R(6,1)$, and $S(4,-14)$ **14.**_____

15. $P(-5,-9)$, $Q(-4,5)$, $R(-9,-7)$, and $S(-23,-6)$ **15.**_____

Objective 4 Interpret slopes in applied problems.

Solve.

16. What is the slope of this roof?   **16.**_____

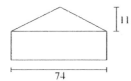

17. The total number of miles of bike paths in a certain   **17.**_____
county over several years is shown in the line
graph. Is the total number of miles of bike paths
increasing or decreasing?

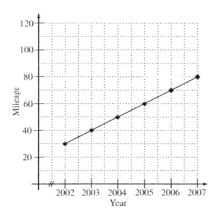

18. The graph shows the amount of garbage (in tons) deposited in landfill A and landfill B from 3 to 7 months after each landfill was opened. Determine whether the amount of garbage in each landfill is growing at the same rate.

18._____

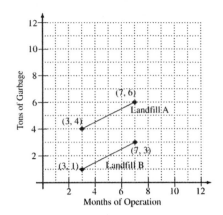

Chapter 3   GRAPHS, LINEAR EQUATIONS AND INEQUALITIES, AND FUNCTIONS

3.3 Graphing Linear Equations

Learning Objectives
1   Identify solutions of linear equations in two variables.
2   Graph a linear equation in two variables.
3   Identify the *x*- and *y*-intercepts of a line.
4   Solve applied problems involving the graph of a linear equation.

Key Terms
Use the most appropriate term or phrase from the given list to complete each statement in exercises 1-4.

x-axis	solution	three points	graph	*y*-intercept
x-intercept	ordered pair	*y*-axis	set	table

1.   The *x*-intercept of a graph is the point where the graph intersects the

_____

2.   The *y*-value of a(n) _____ must be 0.

3.   A(n) _____ of an equation in two variables is an ordered pair of numbers that makes the equation true.

4.   To graph a linear equation in two variables, choose three *x*-values and enter them into a

_____ .

Objective 1 Identify solutions of linear equations in two variables.

Determine if the given ordered pair is a solution of the given equation.

1. $(3,5); y = -\dfrac{4}{9}x + \dfrac{10}{9}$ 1._____

2. $(-9,-5)$; $7x-7y=18$ 2._____

3. $(2,0)$; $9x-3y=18$ 3._____

4. $(5,3)$; $y=3$ 4._____

5. $(-5,5)$; $y=x-\dfrac{3}{7}$ 5._____

Objective 2 Graph a linear equation in two variables.

Graph each equation.

6. $y=-2x+5$ 6.

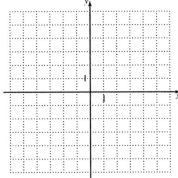

7. $y = -3x - 4$

7.

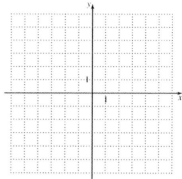

8. $x + 3y = 6$

8.

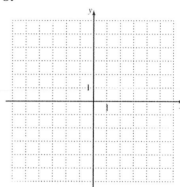

9. $x - y = 6$

9.

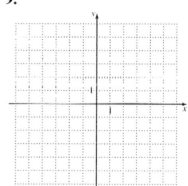

10. $5x - 4y = 20$

10.

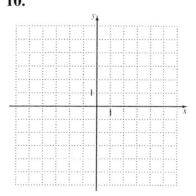

11. $5y - 5x = 15$

11.

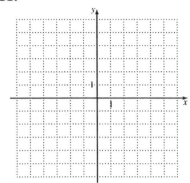

12. $8x + 2y + 4 = 0$

12.

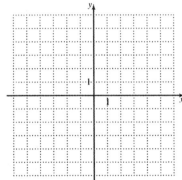

13. $6x + 2y - 8 = 0$

13.

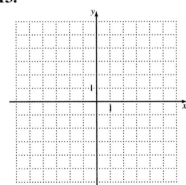

14. $4x = 20$

14.

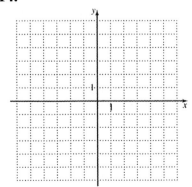

15. $0.5y - 1.5 = 0.5x$ **15.**

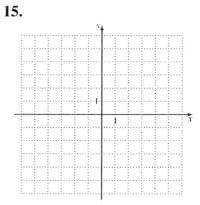

Objective 3   Identify the *x*- and *y*-intercepts of a line.

Find the x- and y-intercepts of the graph of each equation.

16. $y = x - 3$ **16.**_____

17. $5x - 4y = 60$ **17.**_____

18. $7x - 4y = 28$ **18.** _____

19. $-3y = 6$ **19.**_____

20. $-4x = 12$ **20.**_____

Objective 4 Solve applied problems involving the graph of a linear equation.

Solve.

21. A young couple buys an appliance for $740, agreeing to pay $20 down and $30 at the end of each month until the entire debt is paid off.

    a. Express the amount P paid off in terms of the number m of monthly payments.

    b. Complete the table.

    c. Choose an appropriate scale and graph the equation.

21a._____

b.

m	1	2	3
P			

c.

22. Media Services charges $30 for a phone and $20 per month for its economy plan.

    a. Find an equation for the cost, C for t months of phone service.

    b. Choose an appropriate scale and graph the equation.

    c. Find the total cost for 4 months of service.

    d. If a customer has only $80 available, how many months of service can she receive?

22a._____

b.

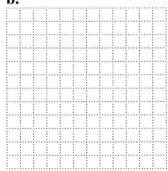

c._____

d._____

Chapter 3   GRAPHS, LINEAR EQUATIONS AND INEQUALITIES, AND FUNCTIONS

3.4 More on Graphing Linear Equations

Learning Objectives
1   Graph a line using its slope and y-intercept.
2   Write a linear equation in slope-intercept form.
3   Write a linear equation in point-slope form.
4   Find the equation of a line, given two points on the line or its slope and one point.
5   Solve applied problems involving the graph of a linear equation.

Key Terms
Use the most appropriate term from the given list to complete each statement in exercises 1-4.

point-slope form	**one point on the line**	**x-intercept**	**slope**
slope-intercept form	**two points on the line**	**y-intercept**	

1.   Given two points on a line, use the _____ to find the equation of the line.

2.   In the slope-intercept form of a line, the intercept part is the _____ of the line.

3.   In the point-slope form of a linear equation, (x_1, y_1) represents _____ .

4.   The form of a linear equation in which y is isolated is the _____ of the line.

Objective 1 Graph a line using its slope and y-intercept.

Graph the following equations using the slope and y-intercept.
1. $y = -3x + 1$ 1.

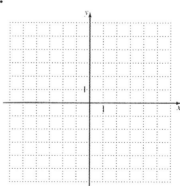

2. $y = 2x - 4$

2.

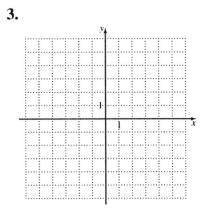

3. $y = -\dfrac{1}{3}x + 5$

3.

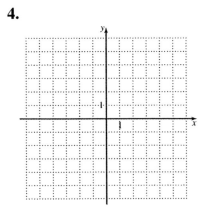

4. $2y + 5x = 4$

4.

5. $y = \dfrac{1}{5}x + 2$

5.

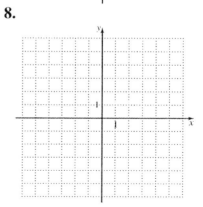

6. $6y + 7x = 6$

6.

7. $7x + 4y = 0$

7.

8. $y = -2.5x - 0.5$

8.

Objective 2 Write a linear equation in slope-intercept form.

Write each equation in slope-intercept form. Then identify the slope and y-intercept.

9. $y - 3x = 0$ **9.**_____

10. $5y + 2x = 5$ **10.**_____

11. $6x + 5y = -10$ **11.**_____

12. $2x - 6y = 6$ **12.**_____

13. $2y - x = 5$ **13.**_____

14. $y - 3 = 6(x + 1)$ **14.**_____

15. $y + 1 = 2(x - 7)$ **15.**_____

Objective 3   Write a linear equation in point-slope form.

Write an equation for each line in point-slope form.

16.

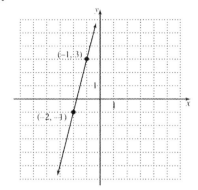

16.

17.

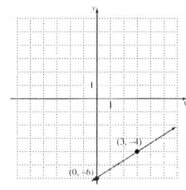

17._____

Objective 4   Find the equation of a line, given two points on the line or its slope and one point.

Write an equation for each line in point-slope form using the information given.

18. slope: $\dfrac{3}{4}$; *y*-intercept: $(0,5)$

18.

19. parallel to the line $y = \dfrac{1}{2}x + 17$ and passes through the point $(4,9)$

19._____

20. perpendicular to the line $3x - 5y + 7 = 0$ and passes through the point $(-3, 3)$

20._____

21. slope: $-\dfrac{5}{3}$; passes through the point $\left(\dfrac{1}{3}, \dfrac{2}{3}\right)$

21._____

22. passes through the points $(7, -1)$ and $(-4, 5)$

22._____

Objective 5 Solve applied problems involving the graph of a linear equation.

Solve.

23. Kara's Custom Tees experienced fixed costs of
$300 and variable costs of $4 per shirt.
   a. Write an equation that can be used to determine
       the total expenses C for producing x shirts.
   b. Find the cost of producing 25 shirts.
   c. Graph the equation.

23a._____

b._____

c.

24. The relationship between the number of pages P
that Jim reads and the number of hours h that he
spends reading is linear. One day he reads 92 pages
in 4 hours, and the next day he reads 23 pages in 1
hour. Using two ordered pairs in the form (h, P),
express the relationship between P and h first in
point-slope form and then in slope-intercept form.

24._____

_____

Name: Date:
Instructor: Section:

Chapter 3   GRAPHS, LINEAR EQUATIONS AND INEQUALITIES, AND FUNCTIONS

3.5 Graphing Linear Inequalities

Learning Objectives
1   Graph a linear inequality in two variables.
2   Solve applied problems involving the graph of a linear inequality.

Key Terms
Use the most appropriate term or phrase from the given list to complete each statement in exercises
1-4.

solid	**linear inequality**	**is not**	**is**
broken	**linear equation**	**test point**	

1.   To determine which side of a boundary lie to shade, use a _____ not on the boundary line.

2.   A _____ line should be used to graph a strict inequality.

3.   An inequality of the form $Ax + By < C$ is a _____.

4.   In the graph of a strict inequality, the boundary line _____ part of the graph.

Objective 1   Graph a linear inequality in two variables.

Determine if the given point is a solution of the given inequality.

1. $y < -4x - 6;\ (-6,5)$ 1._____

2. $y < -9x - 3;\ (-2,4)$ 2._____

3. $7y - 8x > 2;\ (8,15)$ 3._____

4. $4y - 5x \leq 9;\ (5,13)$ 4._____

5. $2y - 3x \leq 2;\ (6,15)$ 5._____

Graph each inequality.

6. $3x + 4y \leq 12$ 6.

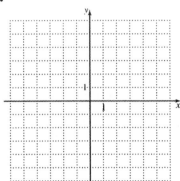

7. $4x + 5y \leq 20$ 7.

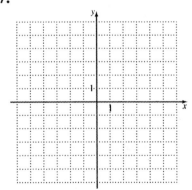

8. $y \leq \dfrac{1}{7}x$ 8.

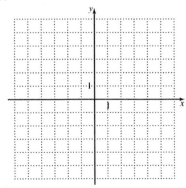

9. $y \le \dfrac{1}{3}x + 2$

9.

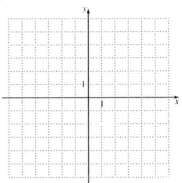

10. $x < 4$

10.

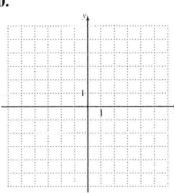

11. $y \le 4$

11.

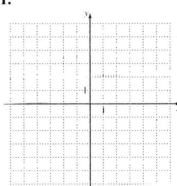

12. $4x - 5y > 20$

12.

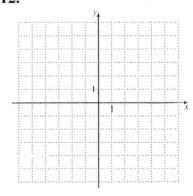

13. $y \le \dfrac{1}{9}x + 2$

13.

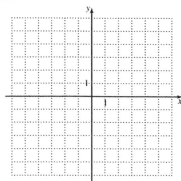

14. $2x - 3y < 6$

14.

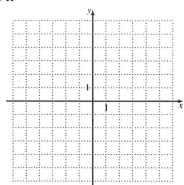

15. $3x - 4y > 12$

15.

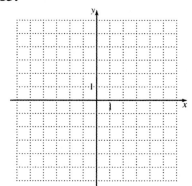

16. $-3x - 2y \ge 6$

16.

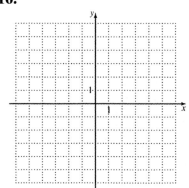

17. $-5x - 6y < 30$

17.

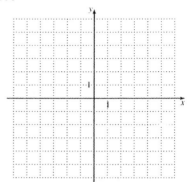

18. $6y - 3x - 18 > 0$

18.

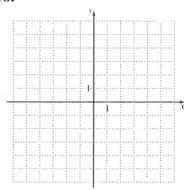

19. $4x + 6y + 24 < 0$

19.

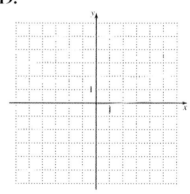

Objective 2 Solve applied problems involving the graph of a linear inequality.

Solve.

20. Wylie sells flowers for $7 apiece and vases for $18   20a._____
     apiece.
     a.   Write an inequality that shows the possible   b.
          combinations of *x* flowers and *y* vases that
          Wylie needs to sell in order to earn at least
          $126.
     b.   Graph the inequality.

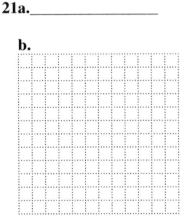

21. Cole washes dogs on the weekends. He charges $5   21a._____
     for small dogs and $8 for large dogs.
     a.   Write an inequality that shows the possible   b.
          combinations of *x* small dogs and *y* large dogs
          that Cole needs to wash in order to earn more
          than $50.
     b.   Graph the inequality.

Name: Date:
Instructor: Section:

Chapter 3   GRAPHS, LINEAR EQUATIONS AND INEQUALITIES, AND FUNCTIONS

3.6 Introduction to Functions

Learning Objectives
1   Identify a function.
2   Determine the domain and range of a function.
3   Evaluate a function written in function notation.
4   Identify various types of functions.
5   Recognize the graph of a function.
6   Solve applied problems involving functions.

Key Terms
Use the most appropriate term or phrase from the given list to complete each statement in exercises 1-6.

output	range	represents	vertical line	does not represent
function	dependent	independent	horizontal line	domain
relation	input			

1.   A function is a _____.

2.   For a particular graph, if no vertical line can be drawn that intersects the graph in more than one point, then the graph _____ a function.

3.   A relation in which no two ordered pairs have the same first coordinates is called a _____.

4.   The _____ of a function is the set of all first coordinates of the ordered pairs that make up the function.

5.   The _____ of a function is the set of all second coordinates of the ordered pairs that make up the function.

6.   For any value of the independent variable, if there is more than one value of the _____ variable then the relation is not a function.

73

Name: Date:
Instructor: Section:

Objective 1   Identify a function.

Determine whether each relation represents a function.

1. $\{(2,5),(4,5),(5,4)\}$ 1._____

2.   2._____

x	-1	1	2	3
y	-3	-4	-2	-3

3.   3._____

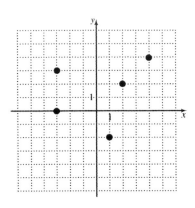

4.   4._____

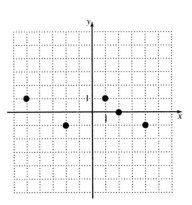

Objective 2   Determine the domain and range of a function.

Find the domain and range of the following functions.

5. $\{(-4,-64),(-2,-8),(0,0),(2,8),(3,27)\}$ 5._____

6.

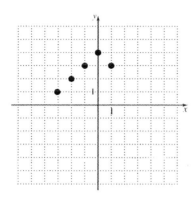

6._____

7.

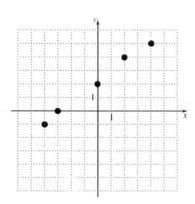

7._____

8.

x	−7	−3	−1	0
y	4	5	2	−2

8._____

Objective 3 Evaluate a function written in function notation.

Evaluate each function for the given values.

9. $f(x) = 3x - 5$

    a. $f(-1)$

    b. $f(3)$

    c. $f\left(\dfrac{1}{3}\right)$

    d. $f(1.2)$

9a._____

b._____

c._____

d._____

10. $g(x) = \dfrac{1}{5}x + 7$

   a. $g(3)$

   b. $g(-5)$

   c. $g\left(\dfrac{5}{4}\right)$

   d. $g(3.5)$

10a._____

   b._____

   c._____

   d._____

11. $f(x) = |x - 9|$

   a. $f(0)$

   b. $f(9)$

   c. $f(t)$

   d. $f(t-1)$

11a._____

   b._____

   c._____

   d._____

12. $h(x) = x^2 + 3x - 1$

   a. $h(2)$

   b. $h(-3)$

   c. $h(z)$

   d. $h(-2z)$

12a._____

   b._____

   c._____

   d._____

13. $f(x) = -7$

   a. $f(-4)$

   b. $f(13)$

   c. $f(-7n)$

   d. $f(x^2)$

13a._____

   b._____

   c._____

   d._____

Objective 4 Identify various types of functions.

Graph each function. Then identify its domain and range.

14. $f(x) = 4x + 2$

14._____

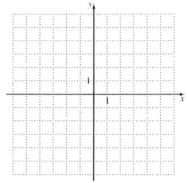

15. $f(x) = 3x - 1$

15._____

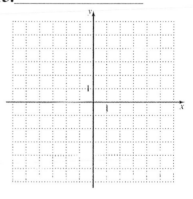

16. $f(x) = |x| + 4$

16._____ _____

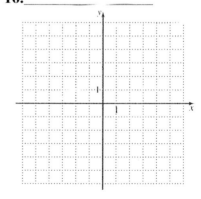

17. $f(x) = -\frac{1}{5}x - 2$

Objective 5 Recognize the graph of a function.

Determine whether each graph represents a function.

18.

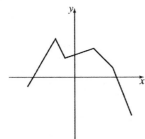

18._____

19.

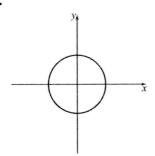

19._____

20.

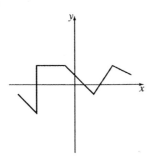

20._____

21.

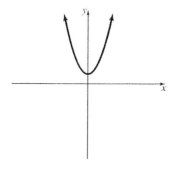

21._____

Objective 6 Solve applied problems involving functions.

Solve.

22. The Fax Mailer bought a fax machine for $600. The
fax machine depreciates at a rate of $20 a month.
   a. Find a function F that can be used to determine
      the value of the fax machine t months after
      purchase.
   b. Graph the function.
   c. Find the domain of F.

22a._____

b.

c._____

23. Media Services charges $40 for a phone and $20 per
month for its economy plan.

    a. Formulate a model that can be used to
determine the total cost, $C(t)$, of operating a
Media Services phone for t months.

    b. Graph the model.

    c. Find the total cost for 10 months of service.

    d. Find the domain of C.

23a. _____

b.

c. _____

d. _____

Chapter 4   SYSTEMS OF LINEAR EQUATIONS AND INEQUALITIES

4.1 Solving Systems of Linear Equations by Graphing

Learning Objectives
1   Determine whether an ordered pair is a solution of a system of linear equations.
2   Solve a system of linear equations by graphing.
3   Determine the number of solutions of a system of linear equations.
4   Solve applied problems involving systems of linear equations by graphing.

Key Terms
Use the most appropriate term or phrase from the given list to complete each statement in exercises 1-4.

inconsistent	set	true	false	infinitely many
independent	no	graph	one	dependent

1.   In a system of equations, if any solution of one equation is also a solution of all other equations in the system, then the system is said to be _____ _____.

2.   A solution of a system of equations in two variables is an ordered pair of numbers that makes both equations in the system _____.

3.   You can tell how many solutions there are for a system of equations by looking at its _____.

4.   If the graphs of a set system of equations intersect in exactly one point, then we say that the equations are _____.

Objective 1   Determine whether an ordered pair is a solution of a system of linear equations.

Determine whether the ordered pair is a solution of the system of equations.

1. $9x - y = 60$
     $4x - 4y = 16$ $(7,3)$

1._____

2. $\begin{array}{l} 3x - y = 31 \\ 8x - 4y = 56 \end{array}$ $(9,4)$

2._____

3. $\begin{array}{l} 3x + 5y = -11 \\ 4x + 2y = -10 \end{array}$ $(-2,-1)$

3._____

4. $\begin{array}{l} 6x - y = 34 \\ 3x - 3y = -3 \end{array}$ $(7,8)$

4._____

Objective 2 Solve a system of linear equations by graphing.
Objective 3 Determine the number of solutions of a system of linear equations.

Solve by graphing.

5. $\begin{array}{l} y = 6x - 8 \\ y = \dfrac{1}{2}x + 3 \end{array}$

5.

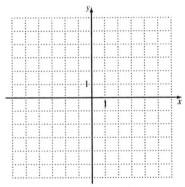

6. $\begin{array}{l} y = 2x - 4 \\ y = \dfrac{1}{2}x - 1 \end{array}$

6.

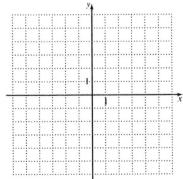

7. $x = y - 3$
$4x = 8y - 16$

7.

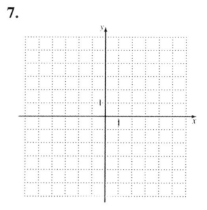

8. $y = -2x + 2$
$5x + y = 5$

8.

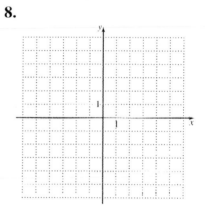

9. $y = 3x - 3$
$6x + y = 6$

9.

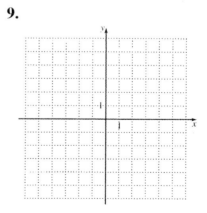

10. $3x - 9y = 54$
$2x - 6y = -12$

10.

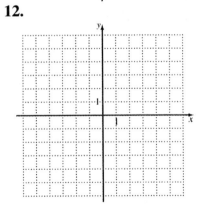

11. $6x - 9y = -27$
$-2x + 3y = 9$

11.

12. $3x - 2y = 6$
$3x - y = 12$

12.

Solve the system using a grapher.

13. $3x + y = 2$
$2x - y = -11$

13._____

14. $5x - 2y = 7$
$y = -4x + 3$

14._____

Objective 4 Solve applied problems involving systems of linear equations by graphing.

Solve.

15. A married couple has a combined annual income of
    $67,000. The wife makes $3000 more than her
    husband.
    a. Express the given information as a system of
        equations.
    b. Graph the system.
    c. What is each of their incomes?

    15a._____

    b.
    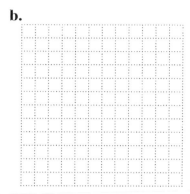

    c._____

16. Tom's Tree and Landscaping Service charges $200
    for a consultation fee plus $50 per hour for labor.
    Lawn Perfect Landscape Service charges $320 for a
    consultation fee plus $30 per hour for labor.
    a. Express the given information as a system of
        equations.
    b. Graph the system.
    c. How many hours of landscaping must be used
        for both services to have the same cost?

    16a.____   __

    b.

    c._____

17. One evening 1600 concert tickets were sold for the
Fairmont Summer Jazz Festival. Tickets cost $25
for covered pavilion seats and $20 for lawn seats.
Total receipts were $36,000.
    a. Express the given information as a system of
        equations.
    b. Graph the system.
    c. How many of each type of tickets were sold?

17a._____

b.

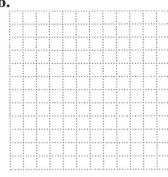

c._____

18. A small company duplicates DVDs. The cost of
duplicating is $39 fixed overhead plus $0.20 per DVD
duplicated. The company generates revenues of $1.50
per DVD. Determine the break-even point for
duplicating DVDs.

18._____

Chapter 4   SYSTEMS OF LINEAR EQUATIONS AND INEQUALITIES

4.2 Solving Systems of Linear Equations Algebraically by Substitution or Elimination

---

Learning Objectives
1 Solve a system of linear equations by substitution.
2 Solve a system of linear equations by elimination.
3 Solve applied problems involving systems of linear equations using substitution or elimination.

Key Terms
Use the most appropriate term or phrase from the given list to complete each statement in exercises 1-4.

no solution	**addition**	**solution**	**simplified**
elimination	**substitution**	**reciprocals**	**original**
both variables	**opposites**	**one variable**	**true** **false**

1. When solving a system of linear equations algebraically, the system has infinitely many solutions if the result is a(n) _____ statement.

2. When solving a system of linear equations algebraically, the system has no solution if the result is a(n) _____ statement.

3. When using elimination or substitution to solve a system of equations, you should always check your solutions by substituting the values in the equations of the _____ system.

4. A system of linear equations in which one equation is solved for one of the variables is especially suitable for using the _____ method to solve the system.

Objective 1 Solve a system of linear equations by substitution.

Solve by substitution.

1. $y = x - 4$
        $y = -x + 8$

1._____

2. $x = 7y - 9$

   $x + 7y = 2$

2._____

3. $9x + 36y = 27$

   $x = 3 - 4y$

3._____

4. $8x + 7y = -42$

   $-7x + y = 51$

4._____

5. $12x - 3y = 9$

   $4x = y - 3$

5._____

6.
$$3x + 5y = 4$$
$$-8x + y = 18$$

6._____

Objective 2 Solve a system of linear equations by elimination.

Solve by elimination.

7.
$$x + 7y = 23$$
$$-x + 3y = -3$$

7._____

8.
$$3x - 2y = 5$$
$$-6x + 4y = -10$$

8._____

9.
$$2x - 4y = 1$$
$$2x - 4y = 2$$

9._____

10. $\dfrac{1}{3}x + \dfrac{1}{2}y = \dfrac{1}{2}$

$\dfrac{1}{3}x + 3y = 8$

10._____

11. $2x - 5y = -14$
$5x + 2y = 52$

11._____

12. $5r - 7s = -51$
$7r + 5s = 47$

12._____

Objective 3 Solve applied problems involving systems of linear equations using substitution or elimination.

Solve.

13. A train leaves a station and travels north at 60 km/hr. Two hours later, a second train leaves the same station on a parallel track and travels north at 100 km/hr. How far from the station will they meet?

13._____

90

14. The perimeter of a standard-sized rectangular rug is 36 ft. The length is 2 ft longer than the width. Find the dimensions.

14._____

15. A nontoxic floor wax can be made from lemon juice and a food-grade linseed oil. The amount of oil should be twice the amount of lemon juice. How much of each ingredient is needed to make 42 ounces of floor wax?

15._____

16. The Jurassic Zoo charges $13 for each adult admission and $5 for each child. The total bill for the 140 people from a school trip was $892. How many adults and how many children went to the zoo?

16._____

17. One canned juice drink is 20% orange juice;
another is 5% orange juice. How many liters of
each should be mixed together in order to get 15 L
that is 16% orange juice?

17._____

Chapter 4   SYSTEMS OF LINEAR EQUATIONS AND
INEQUALITIES

4.3 Solving Systems of Linear Equations in Three Variables

Learning Objectives
1   Determine whether an ordered triple is a solution of a system of linear equations.
2   Solve systems of linear equations in three variables.
3   Solve applied problems involving systems of linear equations in three variables.

Objective 1 Determine whether an ordered triple is a solution of a system of linear equations.

Determine if the given ordered triple is a solution of the given system.

1.
$$x + y + z = -7$$
$$x - 2y - z = -3 \quad (4, 2, -5)$$
$$3x + 2y - z = -3$$

1._____

2.
$$x + y + z = -5$$
$$x - 2y - z = -2 \quad (-6, -5, 6)$$
$$3x + 3y - z = -27$$

2._____

3.
$$x + y + z = -12$$
$$x - 3y - z = 20 \quad (-1, -5, -6)$$
$$2x + 3y - z = -11$$

3._____

Objective 2 Solve systems of linear equations in three variables.

Solve by the elimination method.

$$
\begin{aligned}
x + y - z &= 8 \\
\textbf{4.} \quad -x + 5y + 2z &= 50 \\
4z &= 16
\end{aligned}
$$

4._____

$$
\begin{aligned}
7r - s + t &= 10 \\
\textbf{5.} \quad 3r + 2s - 3t &= -25 \\
r - 3s + 2t &= 25
\end{aligned}
$$

5._____

$$
\begin{aligned}
6r - s + t &= -26 \\
\textbf{6.} \quad 4r + 2s - 3t &= -13 \\
r - 3s + 2t &= -11
\end{aligned}
$$

6._____

$$
\begin{aligned}
4a + 7b &= -48 \\
\textbf{7.} \quad 8a + 4c &= -28 \\
6b + 4c &= -12
\end{aligned}
$$

7._____

$$x + y + z = 1$$
8. $\quad -x + 2y + z = 2$
$$2x - y = -1$$

8. _____

$$5x + 3y + z = -22$$
9. $\quad x - 3y + 2z = 10$
$$14x - 2y + 3z = -50$$

9. _____

$$x + z = 0$$
10. $x + y + 2z = 9$
$$y + z = 4$$

10. _____

$$4x + 3y + z = 1$$
11. $\quad x - 3y + 2z = -31$
$$11x - 2y + 3z = -60$$

11. _____

**Objective 3   Solve applied problems involving systems of linear equations in three
            variables.**

Solve.

12. A hardware supplier manufactures three kinds of   12._____
     clamps, types A, B, and C. Production restrictions
     force it to make 20 more type C clamps than the
     total of the other types and twice as many type B
     clamps as type A clamps. The shop must produce
     260 clamps per day. How many of each type are
     made per day?

13. The perimeter of a triangle is 47 cm. The longest   13._____
     side is 1 cm shorter than the sum of the other two
     sides. Twice the shortest side is 13 cm less than the
     longest side. Find the length of each side of the
     triangle.

14. A basketball team recently scored a total of 99   14._____
     points on a combination of 2-point field goals, 3-
     point field goals, and 1-point foul shots. Altogether,
     the team made 52 baskets and 21 more 2-pointers
     than foul shots. How many shots of each kind were
     made?

15. An investment of $86,000 was made by a business   15._____
     club. The investment was split into three parts and
     lasted for one year. The first part of the investment
     earned 8% interest, the second 6%, and the third
     9%. Total interest from the investments was $6840.
     The interest from the first investment was 6 times
     the interest from the second investment. Find the
     amounts of the three parts of the investment.

Chapter 4   SYSTEMS OF LINEAR EQUATIONS AND INEQUALITIES

4.4 Solving Systems of Linear Equations by Using Matrices

Learning Objectives
1   Solve systems of two or three linear equations using matrices.
2   Solve applied problems involving systems of linear equations using matrices.

Key Terms

Use the most appropriate term from the given list to complete each statement in exercises 1-4.

constants	square	coefficients	adding the products to
numeral	columns	equivalent	multiplying the result with
element	rows	augmented	matrix

1.   In an augmented matrix, the _____ are on the right side of the vertical bar.

2.   The _____ of a matrix are vertical.

3.   To solve a system of linear equations using matrices, first write the corresponding _____ matrix for the system.

4.   If a square matrix has three columns then it must have three _____ .

Objective 1 Solve systems of two or three linear equations using matrices.

Solve each system using matrices.

1. $\begin{aligned} x + y &= 2 \\ x - y &= 10 \end{aligned}$ 1._____

2. $\begin{aligned} 9x - 5y &= 61 \\ 4x + y &= 11 \end{aligned}$ 2._____

3.
$$7x - 5y = 31$$
$$2x + y = 4$$

3._____

4.
$$5x + 2y = -5$$
$$-5x - 2y = -9$$

4._____

5.
$$x + y - z = 1$$
$$3x - y + z = -9$$
$$x - 2y + 4z = 12$$

5._____

6.
$$x + y - z = -6$$
$$2x - y + z = 0$$
$$x - 4y + 3z = -5$$

6._____

$$-4x - 3y = -2$$

7. $12x + 9y = -9$

$$5x - 6y = -7$$

7._____

$$7x - y - 5z = 5$$

8. $6x + y - z = 7$

$$3x + y - 7z = 7$$

8._____

$$5x - 6y + 6z = -1$$

9. $-10x + 12y - 12z = 2$

$$-15x + 18y - 18z = 3$$

9._____

$$7x - y = 6$$

10. $y - z = 7$

$$5x + y = 3$$

10._____

Objective 2 Solve applied problems involving systems of linear equations using matrices.

Solve.

11. A grocer mixes candy worth $0.80 per pound and nuts worth $0.70 per pound to get an 18-pound mixture worth $0.77 per pound. How many pounds of candy should be used?

11._____

12. Fran receives $113 per year in simple interest from investing $1300. Part is invested at 7%, part at 8%, and part at 9%. There is $800 more invested at 9% than at 8%. Find the amount invested in each account.

12._____

Chapter 4   SYSTEMS OF LINEAR EQUATIONS AND INEQUALITIES

4.5 Solving Systems of Linear Inequalities

Learning Objectives
1   Solve a system of linear inequalities by graphing.
2   Solve applied problems involving systems of linear inequalities.

Key Terms

Use the most appropriate term or phrase from the given list to complete each statement in exercises 1-2.

shaded regions	triples	boundary line	pairs
simultaneous	both	coordinate plane	a system of

1. A solution to a system of two linear inequalities is a point that lies in both _____ of the graph.

2. The graph of the inequality $2x + 3y \leq -1$ includes the _____.

Objective 1   Solve a system of linear inequalities by graphing.

Solve by graphing.

1.
$$y < 8x - 3$$
$$y < -4x + 5$$

1.

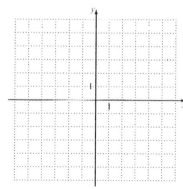

Name:

Instructor:

Date:

Section:

2.
$$y > 4x - 3$$
$$y < -2x + 5$$

2.

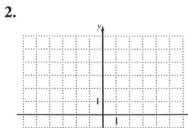

3.
$$x - 3y \geq 9$$
$$3x + y < 9$$

3.

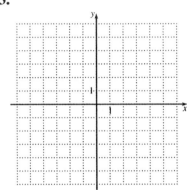

4.
$$x - 2y \geq 4$$
$$2x + y \leq 4$$

4.

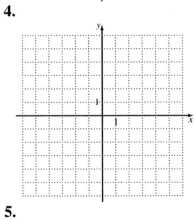

5.
$$5x + 2y < 10$$
$$5x - 3y < 15$$

5.

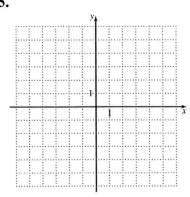

6. $3x + 3y < 9$
       $2x - 2y < 4$

6.
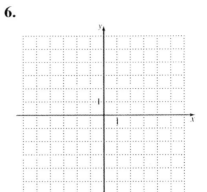

7. $y < 3x + 1$
       $3x - y \leq 4$

7.
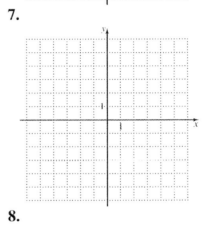

8. $y < 2x + 1$
       $2x - y \leq 6$

8.
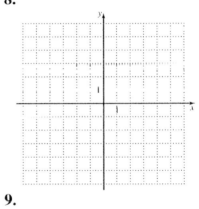

9. $y > -3$
       $x \leq 4$

9.
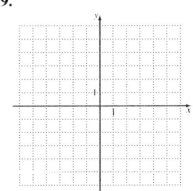

10. $y \le -\dfrac{2}{3}x - 4$

$y > 3x + 1$

10.

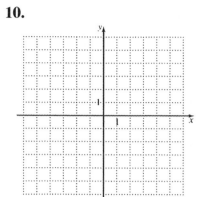

11. $y < 1.5x - 3$

$y > -0.5x + 2$

11.

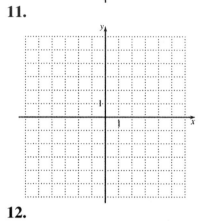

12. $4x + 5y \le 10$

$y \ge -2$

$x \ge -5$

12.

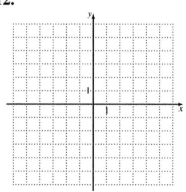

13. $y \le 2x - 3$

$y \ge -2x - 3$

$x \le 2$

13.

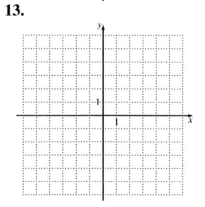

14. $\begin{aligned} y &\leq 3x+3 \\ y &> -3x+3 \\ x-2 &< 0 \end{aligned}$

14.

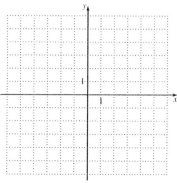

15. $\begin{aligned} y &\leq 2x+4 \\ y &> -2x-1 \\ x-2 &< 0 \end{aligned}$

15.

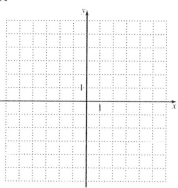

Objective 2 Solve applied problems involving systems of linear inequalities.

Solve.

16. A college student works in both the school cafeteria and library. She works no more than 10 hours per week in the cafeteria and no more than 17 hours per week in the library. She must work at least 20 hours each week.
   a. Express this information as a system of inequalities.
   b. Graph the system.
   c. How many hours can she work in the library if she works 8 hours in the cafeteria in one week?

16a._____

b.

c._____

17. Carlo and Anita make mailboxes and toys in a craft shop. Each mailbox, x, requires 1 hour of work from Carlo and 1 hour from Anita. Each toy, y, requires 1 hour of work from Carlo and 3 hours from Anita. Carlo can work no more than 7 hours per week and Anita can work no more than 15 hours per week.

    a. Express this information as a system of inequalities.

    b. Graph the system.

    c. What does the solution region represent?

17a._____

b.

c._____

Name: Date:
Instructor: Section:

Chapter 5   POLYNOMIALS

5.1 Addition and Subtraction of Polynomials

Learning Objectives
1   Classify polynomials.
2   Simplify polynomials.
3   Evaluate polynomials.
4   Add and subtract polynomials.
5   Solve applied problems involving polynomials.

Key Terms
Use the most appropriate term or phrase from the given list to complete each statement in exercises 1-4.

multiplied or divided	**polynomial**	**monomial**	**leading term**
with two terms	**variables**	**terms**	**lowest**
leading coefficient	**missing term**	**constant term**	**binomials**
added or subtracted	**highest**	**of degree two**	

1. A(n) _____ is an algebraic expression with one or more monomials added or subtracted.

2. A binomial is a polynomial _____.

3. The coefficient of the leading term is called the _____ .

4. A(n) _____ function is a function in which the rule that defines the function is a polynomial.

Objective 1   Classify polynomials.

Determine whether each of the following expressions is a polynomial.

1. $4x+7$ 1._____

2. $\frac{5}{n}+n^2-1$ 2._____

3. $\dfrac{x-3}{x+5}$

3._____

4. $3b^2 + 5b - 7$

4._____

Identify the terms and coefficients of each polynomial.

5. $6x^3 - 3x^2 + 4x - 5$

5._____

_____

6. $5x + \dfrac{x^2}{4}$

6._____

_____

7. $2x^2 y^3 - 3x^3 y^2$

7._____

_____

Classify each of the following polynomials by the number of terms it has. Then identify the degree of the polynomial.

8. $8x^2 y^3 z$

8._____

9. $5a^3 b - 7a^2 b^2 + 4ab^3 - 2$

9._____

10. $4s^2 t - st + 2st^2$

10._____

11. $7x^2 y^4 z^3 - 8x^4 y^3 z$

11._____

Rewrite each polynomial in descending order. Then identify the leading term and leading coefficient.

12. $4p^3 + 5p - 3p^4 + 7 + 2p^2$ 12._____

13. $6y^2 - 7y^4 + 7y - 3$ 13._____

14. $5a^5 + 3a^2 - a^3 + a^4 + 1 - a$ 14._____

Objective 2 Simplify polynomials.

Simplify, and then write in descending order.

15. $9b^5 + b^2 - b^3 - 3b^5 - 5b^2$ 15._____

16. $-x + \dfrac{5}{8} + 38x^6 - x - \dfrac{1}{2} - 2x^6$ 16._____

Objective 3 Evaluate polynomials.

Evaluate each polynomial for the given value of the variable.

17. $2x^2 - 4x + 7;\ x = 5$ 17._____

18. $-5x^3 + 7x^2 - 7x - 2;\ x = 0$ 18._____

Objective 4 Add and subtract polynomials.

Add or subtract.

19. $\left(-3x + 3\right) + \left(x^2 + x - 6\right)$ 19._____

20. $\left(a^2 - 3a + 2\right) + \left(-2a^2 + 4a - 4\right)$ 20._____

21. $\left(7x^4+9x^3-6\right)-\left(7x^2-8x+8\right)$ 21._____

22. $\left(9x^5-2x^4-2x\right)-\left(5x^3+6x^4+x\right)$ 22._____

23. Add $7f^3-6f^2-9$ and $8f^3+3f^2+3f$.   23._____

24. Subtract $7r^5-5r^2+12$ from $3r^5+7r^2-4$.   24._____

Simplify.

25. $\left(3a^2-2a+4\right)-\left(7a^2+4a-3\right)+\left(4a^2-3a+1\right)$ 25._____

26. $\left(5y^3+14\right)+\left(5y^3+6y-3\right)-\left(6y^3+y^2-2y\right)$ 26._____

Given the functions f(x) and g(x), find f(x) + g(x) and f(x) – g(x).

27. $f\left(x\right)=4x^2+7x-5$ and $g\left(x\right)=-9x^2+3x-12$ 27._____

28. $f\left(x\right)=3x^3+5x^2-7$ and $g\left(x\right)=-8x^3-4x^2-9$ 28._____

Objective 5 Solve applied problems involving polynomials.

Solve.

29. Total revenue is the total amount of money taken in   29._____
by a business. An appliance firm determines that
when it sells x washing machines, the total revenue,
R, in dollars is given by the polynomial
$R=272.04x-0.2x^2$. What is the total revenue
from the sale of 146 washing machines?

Chapter 5   POLYNOMIALS

5.2 Multiplication of Polynomials

Learning Objectives
1   Multiply polynomials.
2   Multiply binomials using FOIL.
3   Use a special product formula to square a binomial or to multiply the sum and difference of two binomials.
4   Solve applied problems involving the multiplication of polynomials.

Key Terms
Use the most appropriate term or phrase from the given list to complete each statement in exercises 1-4.

distributive property	square of a binomial	product rule of exponents
FOIL method	binomials	monomials
negative	positive	opposite outer

1.   To multiply a polynomial by a monomial, use the _____.

2.   In FOIL, the 'O' stands for _____.

3.   The formula $(a-b)^2 = a^2 - 2ab + b^2$ is called the _____.

4.   The FOIL method is used to multiply _____.

Objective 1 Multiply polynomials.

Multiply.

1. $\left(4x^9\right)\left(4x^9\right)$ 1._____

2. $\left(7x^5\right)\left(6x^3\right)$ 2._____

3. $\left(2st\right)\left(8s^2t^5\right)\left(4s^3\right)$ 3._____

4. $(6qr)(8q^3r^3)(2q^5)$ 4._____

Find each product.

5. $8x(-x+3)$ 5._____

6. $2x(3x^2-8x+8)$ 6._____

7. $4x(8x^3-3x^2+7x)$ 7._____

8. $9x^2y(9x^2-6xy+2y^2)$ 8._____

Simplify.

9. $5(2x^2-7x+30)+7(4x^2+9x-86)$ 9._____

10. $8x^2(7x^2-7x+20)+3x^2(8x^2-9x-72)$ 10._____

Objective 2 Multiply binomials using FOIL.

Multiply.

11. $(x+5)(x+4)$ 11._____

12. $(x-9)(x-4)$

12._____

13. $(b+2)^2$

13._____

14. $(2a+8)(a+4)$

14._____

15. $(8z+2)(9z+4)$

15._____

16. $(5x+7)^2$

16._____

17. $(8z-u)(2z+4u)$

17._____

18. $(6z-u)(2z+7u)$

18._____

Objective 3 Use a special product formula to square a binomial or to multiply the sum and difference of two binomials.

Use a special product formula to simplify.

19. $(x+11)^2$

19._____

20. $(5w+5)(5w-5)$

20._____

21. $(2s+3l)(2s-3l)$

21._____

22._____

22. $\left(v-\dfrac{1}{7}s\right)^2$

Multiply.

23. $(3g-j)^3$

23._____

24. $(r+4)(r-4)(r^2+16)$ 24._____

Given f(x) and g(x), find f(x) · g(x).

25. $f(x)=2x^2-3$ and $g(x)=4x^2-7x+2$ 25._____

26. $f(x)=4x^2-3$ and $g(x)=5x^3-3x^2+x$ 26._____

Evaluate each function for the given expression.

27. $f(x)=9-2x-4x^2; f(m+4)$ 27._____   _____

28. $f(x)=3-8x-5x^2; f(m+p)-f(p)$ 28._____

29. $f(x)=5x+x^2; f(y+h)-f(y)$ 29._____

Name: Date:
Instructor: Section:

Objective 4 Solve applied problems involving the multiplication of polynomials.

Solve.

30. Find a polynomial that represents the area within   **30.**
         the circle of radius *x* but outside of the 5 by 1
         rectangle.

116

Name: Date:
Instructor: Section:

Chapter 5   POLYNOMIALS

5.3 Division of Polynomials

Learning Objectives
1   Divide a polynomial by a monomial.
2   Divide a polynomial by a polynomial.
3   Solve applied problems involving the division of polynomials.

Key Terms
Use the most appropriate term or phrase from the given list to complete each statement in exercises 1-2.

divisor   remainder   missing terms

dividend   quotient   leading terms

1.   In the division problem $\left(-8x^2 + 26x - 19\right) \div \left(2x - 3\right) = \left(-4x + 7\right) + \dfrac{2}{2x - 3}$, $-4x - 7$ is

called the _____.

2.   In the division problem $\left(-8x^2 + 26x - 19\right) \div \left(2x - 3\right) = \left(-4x + 7\right) + \dfrac{2}{2x - 3}$, the

numerator of $\dfrac{2}{2x - 3}$ is called the ____   _____.

Objective 1 Divide a polynomial by a monomial.

Simplify.

1. $\dfrac{-64r^{11}}{-4r^7}$ 1._____

2. $\dfrac{6a^6 b^4}{-2ab}$ 2._____

3. $\dfrac{15s^4 t^3}{-3st}$ 3._____

Name: Date:
Instructor: Section:

Find each quotient.

4. $\left(8x^7 - 24x^5\right) \div (4x)$ 4._____

5. $\left(35x^8 - 63x^3\right) \div (7x)$ 5._____

6. $\dfrac{15x^3 - 12x^2 + 3}{3}$ 6._____

7. $\dfrac{12x^3 - 20x^2 + 16}{4}$ 7._____

8. $\dfrac{12m^3 + 9m^2 + 9m}{3m^2}$ 8._____

9. $\dfrac{45v^9 p^{10} - 20v^7 p^8 + 20v^3 p^4}{5v^3 p}$ 9._____

Objective 2 Divide a polynomial by a polynomial.

Divide.

10. $\left(c^2 + 9c + 14\right) \div (c + 2)$ 10._____

11. $\dfrac{8x^3 - 20x^2 - 32x - 10}{4x + 2}$ 11._____

12. $\dfrac{16x^3 - 12x^2 - 44x - 16}{4x + 4}$

        12._____

13. $\dfrac{28w^3 + 20w^2 + 24w + 39}{4w + 4}$

        13._____

14. $\dfrac{3b^2 + 12b + 9}{3b + 3}$

        14._____

15. $\left(32k^4 + 28k^3 + 7k - 2\right) \div \left(4k^2 + 1\right)$

        15._____

16. $\left(4n^4 + 4n^3 + 2n - 1\right) \div \left(2n^2 + 1\right)$

        16._____

17. $\dfrac{x^3 + 1000}{x + 10}$

        17._____

18. $\dfrac{x^3+27}{x+3}$ 18._____

19. $\dfrac{9k^4+21k^3+7k-1}{3k^2+1}$ 19._____

Given f(x) and g(x), find $\dfrac{f(x)}{g(x)}$.

20. $f(x)=5x^2+22x+21$ and $g(x)=x+3$ 20._____

Objective 3 Solve applied problems involving the division of polynomials.

Solve.

21. A long-distance phone service charges $8.32 per   21a._____
month plus $0.04 per minute for long-distance
phone calls. The average cost per minute of a phone
call (in dollars) is given by the expression
$\dfrac{8.32+0.04x}{x}$, where x is the total number of   b._____
monthly calling minutes.
  a. Use division to rewrite this expression.
  b. Use the expression from part (a) to find the
       average cost per minute, to the nearest cent, if
       1 hour of long distance calls are made in a
       month.

Chapter 5   POLYNOMIALS

5.4 The Greatest Common Factor and Factoring by Grouping

Learning Objectives
1   Factor out the greatest common factor from a polynomial.
2   Factor a polynomial by grouping.
3   Solve applied problems involving factoring.

Key Terms
Use the most appropriate term from the given list to complete each statement in exercises 1-3.

using the GCF	divide	factor	common multiples
common factors	negative	product	distributive property
highest	sum	quotient	that binomial

1.   The greatest common factor of two monomials is the _____ of the greatest common factor of the coefficients and the highest powers of the variables common to all of the monomials.

2.   When the terms of a polynomial have common factors, use the _____ in reverse to factor the polynomial.

3.   To _____ a polynomial is to rewrite it as a product.

Objective 1 Factor out the greatest common factor from a polynomial.

Factor out the greatest common factor.

1. $20y^4 + 5y$ 1._____

2. $3x^7 - 21x^6 + 21x^5$ 2._____

3. $7x^6 - 28x^5 + 14x^4$ 3._____

4. $-2x^6y^4 - 8x^4y^3 - 16xy$ 4._____

5. $-4m^7n^5 - 28m^5n^4 - 28m^2n^2$

5._____

6. $x(y-4)+9(y-4)$

6._____

7. $r(b-5)+7(b-5)$

7._____

8. $t^2(t+2w^2)-2w(t+2w^2)+(t+2w^2)$

8._____

9. $r^2(r+7s^2)-5s(r+7s^2)+(r+7s^2)$

9._____

Solve for the indicated variable.

10. $B = ch + fh$, for h

10._____

Objective 2 Factor a polynomial by grouping.

Factor by grouping.

11. $p^2(y-t)+3(t-y)$

11._____

12. $h^2(x-v)+5(v-x)$

12._____

13. $qs - 8s + qf - 8f$ **13.**_____

14. $wv - 9v + wk - 9k$ **14.**_____

15. $b^2 + 3b + 6b + 18$ **15.**_____

16. $z^2 + 5z + 7z + 35$ **16.**_____

17. $6st + 15t - 10s - 25$ **17.**____   ____

18. $4xy + 14y - 10x - 35$ **18.**_____

19. $r^2 - 10tw + 5wr - 2tr$ **19.**_____

20. $j^2 - 12kl + 4jl - 3jk$ 20._____

21. $2a^3 + 2ab^2 + 5a^2b + 5b^3$ 21._____

22. $3c^3 + 3cd^2 + 7c^2d + 7d^3$ 22._____

Objective 3 Solve applied problems involving factoring.

Solve.

23. An object is thrown upward from ground level with   23._____
an initial velocity of 96 feet per second. Its height
after t seconds is a function h given by
$h(t) = 96t - 16t^2$.

   a. Find an equivalent expression for $h(t)$ by

       factoring out a common factor.

   b. Find the height of the object when $t = 2$.

24. When x hundred DVD players are sold, Rolics   24._____
Electronics collects a profit of $P(x)$, where

   $P(x) = x^2 - 2x$ and $P(x)$ is in thousands of

dollars. Find an equivalent expression by factoring
out the greatest common factor.

Chapter 5   POLYNOMIALS

5.5 Factoring Trinomials

Learning Objectives
1   Factor trinomials of the form $x^2 + bx + c$.
2   Factor trinomials of the form $ax^2 + bx + c$, $a \neq 1$.
3   Solve applied problems involving factoring.

Key Terms

Use the most appropriate term or phrase from the given list to complete each statement in exercises 1-2.

the number b whose sum is ac	**product**	**sum**
the ac method **factors of b**	**binomials**	**prime**
factors of c **multiplication**	**trial-and-error**	**FOIL**

1. Trinomials that cannot be written as the product of two binomials with integer coefficients are called _____ polynomials.

2. An alternative procedure for factoring the trinomial $ax^2 + bx + c$ based on grouping is called _____.

Objective 1 Factor trinomials of the form $x^2 + bx + c$.

Fill in the missing factor.

1. $x^2 - 12x + 27 = (x-3)(\quad)$ 1._____

2. $x^2 + 18x + 72 = (x+6)(\quad)$ 2._____

Factor, if possible.

3. $r^2 + 10r + 21$ 3._____

4. $c^2 + 14c + 48$ 4._____

5. $w^2 - 6w + 8$ 5._____

6. $t^2 - 9t + 20$ 6._____

7. $s^2 + 4s - 21$ 7._____

8. $r^2 + 2r - 35$ 8._____

9. $v^2 - v - 20$ 9._____

10. $c^2 - 8cf + 15f^2$ 10._____

11. $a^2 - 2ay - 80y^2$ 11._____

12. $r^2 - 2rf - 48f^2$ 12._____

13. $m^2 + 7mn + 7n^2$ 13._____

14. $-r^2 - 9r + 22$ 14._____

Objective 2   Factor trinomials of the form $ax^2 + bx + c$, $a \neq 1$.

Fill in the missing factor.

15. $8a^2 + 14ab - 15b^2 = (2a + 5b)(\quad\quad)$ **15.** _____

16. $12c^2 + 32cd - 35d^2 = (2c + 7d)(\quad\quad)$ **16.** _____

Factor, if possible.

17. $56s^2 + 31s + 3$ **17.** _____

18. $12x^2 + 43x + 35$ **18.** _____

19. $5c^2 + 19c - 30$ **19.** _____

20. $4s^2 + 9s - 28$ **20.** _____

21. $20w^2 - 11w - 3$ **21.** _____ _____

22. $30z^2 - 19z - 63$ **22.** _____

23. $18c^2 - 33cw + 14w^2$ **23.** _____

24. $6v^2 + 13vw - 63w^2$ 24._____

25. $56s^2 + 23sb - 63b^2$ 25._____

26. $8w^2 - 30w + 18$ 26._____

27. $b^6 - 4b^3 - 60$ 27._____

28. $v^4 + 19v^2 + 18$ 28._____

29. $-12m^2 + 26m + 56$ 29._____

Objective 3 Solve applied problems involving factoring.

Solve.

30. An object is thrown upward so that its height in   30._____
meters above the ground at time t seconds is
represented by the expression $-5t^2 - 17t + 12$.
Factor this expression.

Chapter 5   POLYNOMIALS

5.6 Special Factoring

Learning Objectives
1   Factor perfect square trinomials.
2   Factor the difference of squares.
3   Factor the sum and difference of cubes.
4   Solve applied problems involving factoring.

Key Terms
Use the most appropriate term from the given list to complete each statement in exercises 1-2.

the difference of two perfect cubes	**the difference**
the sum of squares	**the cube of a binomial**
the difference of squares	**the square of the product**

1.   The formula for factoring _____ of two terms says that the factorization is the sum of the two terms times the difference of the same two terms.

2.   The product $(a-b)(a^2+ab+b^2)$ is the factorization of _____.

Objective 1 Factor perfect square trinomials.

Factor, if possible.

1. $s^2 - 8s + 16$ 1._____

2. $b^2 + 14b + 49$ 2._____

3. $v^2 + 6v + 9$ 3._____

4. $9w^2 + 16 + 24w$ 4._____

5. $4c^2 + 81 + 36c$

5._____

6. $12w + 4w^2 + 9$

6._____

7. $81s^2 - 18sf + f^2$

7._____

8. $v^6 + 8v^3 + 16$

8._____

9. $16s^8 - 40s^4f + 25f^2$

9._____

10. $9v^{12} - 24v^6f + 16f^2$

10._____

Objective 2 Factor the difference of squares.

Factor, if possible.

11. $v^2 - 81$

11._____

12. $w^2 - 25$

12._____

13. $36 - v^2$

13._____

14. $49w^2 - 100$ **14.**_____

15. $64x^2 - 121$ **15.**_____

16. $0.09w^2 - 0.16$ **16.**_____

17. $36r^4 - 25$ **17.**_____

18. $49c^4 - 121$ **18.**_____

19. $p^2(x-v) - 4(x-v)$ **19.**_____

20. $c^2 + 8c + 16 - p^2$ **20.**_____

Objective 3 Factor the sum and difference of cubes.

Factor, if possible.

21. $c^3 + 8$ **21.**_____

22. $r^3 - 64$ 22._____

23. $125x^3 + 8y^3$ 23._____

24. $343c^6 - 512d^6$ 24._____

25. $5a^6 - 1080a^3$ 25._____

Factor. Assume that all exponents are positive.

26. $a^{2n} - 64$ 26._____

Objective 4 Solve applied problems involving factoring.

Solve.

27. A \$17,000 investment grew by an average annual rate of   27._____
return of r. After two years, the value of the investment
in dollars was $17,000 + 34,000r + 17,000r^2$. What is the
factorization of this expression?

Chapter 5   POLYNOMIALS

5.7 Solving Quadratic Equations by Factoring

Learning Objectives
1   Solve quadratic equations by factoring.
2   Solve applied problems using quadratic equations.

Key Terms
Use the most appropriate term from the given list to complete each statement in exercises 1-2.

> **the squares of the legs**   **binomial equation**   **quadratic equation**
>
> **the square of the hypotenuse**   **identity property**   **zero**

1.   The Pythagorean Theorem states that for every right triangle, the sum of
     _____ equals _____.

2.   To solve a quadratic equation by factoring, rewrite the equation in standard from with
     _____ on one side.

Objective 1 Solve quadratic equations by factoring.

Solve.

1. $(s+28)(s-32)=0$ 1._____

2. $(7r+4)(3r-9)=0$ 2._____

3. $(5s+3)(2s-6)=0$ 3._____

4. $(s-3)^2=0$ 4._____

5. $c^2 + 5c = 0$

_____

6. $v^2 + 4v = 0$

_____

7. $5r^2 - 9r = 0$

_____

8. $v^2 + 10v + 21 = 0$

_____

9. $r^2 + 9r + 14 = 0$

_____

10. $a^2 - 11a + 24 = 0$

_____

11. $v^2 - 10v + 16 = 0$

_____

12. $18 + 3c - c^2 = 0$

_____

13. $8 + 2a - a^2 = 0$

13. _____

14. $3c^2 + 16c + 21 = 0$

14. _____

15. $32v^2 - 60v + 27 = 0$

15. _____

16. $t^2 = 25$

16. _____

17. $v^2 = 4$

17. _____

18. $5s^2 = 7s$

18. _____

19. $4p^2 = 9p$ **19.**_____

20. $t^2 - 16 = 0$ **20.**_____

21. $c^2 - 4 = 0$ **21.**_____

22. $b(b-4) = 21$ **22.**_____

23. $t(t-5) = 36$ **23.**_____

24. $(x-3)(x+4) = 30$ **24.**_____

25. $(x-2)(x-1)=6$ **25.**_____

Given f(x) and g(x), find all values of x such that f(x) = g(x).

26. $f(x)=x^2+12x+35;\ g(x)=3$ **26.**_____

27. $f(x)=x^2+16x+61;\ g(x)=1$ **27.**_____

Objective 2 Solve applied problems using quadratic equations.

Solve.

28. The length of the top of a table is 7 m greater than   **28.**_____
the width. The area is 78 m². Find the dimensions of
the table.

29. A projectile is launched upward from a height of
768 ft with an initial velocity of 32 feet per second.
Its height after t sec is given by the equation
$h = -16t^2 + 32t + 768$.

    a. Find the number of seconds until it returns to
the ground.

    b. After how many seconds will the projectile
have a height of 528 ft?

29a._____

    b._____

Chapter 6   RATIONAL EXPRESSIONS AND EQUATIONS

6.1 Multiplication and Division of Rational Expressions

Learning Objectives
1   Identify values for which a rational expression is undefined.
2   Simplify rational expressions.
3   Multiply rational expressions.
4   Divide rational expressions.
5   Solve applied problems involving multiplication and division of rational expressions.

Key Terms
Use the most appropriate term or phrase from the given list to complete each statement in exercises 1-4.

inverses	**rational number**	**rational expression**	**factoring**
undefined	**multiply**	**divide**	**denominator**
opposites	**equivalent**	**common factors**	**multiplying**

1.   Every _____ is a rational expression.

2.   A rational expression is _____ when the denominator is equal to zero.

3.   To simplify a rational expression, factor the numerator and denominator and divide out any _____.

4.   When the terms in the numerator and denominator of a rational expression differ only in sign, they are _____ of each other.

Objective 1   Identify values for which a rational expression is undefined.

Identify the values for which the given rational expression is undefined.

1. $-\dfrac{25}{21y}$ 1._____

2. $\dfrac{x-6}{2}$ 2._____

3. $\dfrac{2x-8}{7x+6}$

3._____

4. $\dfrac{4x+2}{x^2-8x+15}$

4._____

5. $\dfrac{3x+1}{x^2-4x-5}$

5._____

6. $\dfrac{r^3-6r}{r^2-64}$

6._____

Objective 2 Simplify rational expressions.

Determine whether each pair of rational expressions is equivalent.

7. $\dfrac{4m+3}{3m-1}$ and $\dfrac{4m^2+3m}{3m^2-m}$

7._____

8. $\dfrac{2x}{7x-4}$ and $\dfrac{3x}{8x-4}$

8._____

Simplify, if possible.

9. $\dfrac{80w^{10}x^5}{50w^7x^2}$

9._____

10. $\dfrac{48v^9 z^5}{18v^6 z^2}$

10. _____

11. $\dfrac{15x^2 + 30x}{5x^2}$

11. _____

12. $\dfrac{12x^2 + 24x}{3x^2}$

12. _____

13. $\dfrac{6}{6a - 14}$

13. _____

14. $\dfrac{1-v}{v-1}$

14. _____

15. $\dfrac{2t - 8}{t^2 - 16}$

15. _____

16. $\dfrac{x^2 - 6x + 5}{x^2 + 2x - 3}$

16. _____

17. $\dfrac{z^2 - 10z + 16}{z^2 + 2z - 8}$

17._____

18. $\dfrac{x^3 - 64}{x^2 - 16}$

18._____

Objective 3 Multiply rational expressions.

Multiply. Express answers in lowest terms.

19. $\dfrac{x^2}{3y} \cdot \dfrac{9y^3}{x^5}$

19._____

20. $\dfrac{x^2}{4y} \cdot \dfrac{16y^5}{x^4}$

20._____

21. $\dfrac{z^2 - z - 30}{5z^3 - 4z^2} \cdot \dfrac{25z^3 - 16z}{6z - 36}$

21._____

22. $\dfrac{z^2 - 3z - 28}{6z^3 - 7z^2} \cdot \dfrac{36z^3 - 49z}{7z - 49}$

22._____

23. $\dfrac{t^2+2t-63}{t^2+3t-70} \cdot \dfrac{t^2-8t-20}{t^2+11t+18}$ 23._____

24. $\dfrac{8r^2-79r+63}{8r^2+57r-56} \cdot \dfrac{9r^2+65r-56}{72r^2-119r+49}$ 24._____

Objective 4 Divide rational expressions.

Divide. Express answers in lowest terms.

25. $\dfrac{11a^4b^5}{9a^3b} \div \dfrac{121a^4b}{231a^6b^3}$ 25._____

26. $\dfrac{z^2-36}{z^4} \div \dfrac{z^6+6z^3}{z+4}$ 26._____

27. $\dfrac{x^2-25}{x^2-10x+25} \div \dfrac{4x-20}{x^2-2x-15}$ 27._____

28. $\dfrac{12x^2+7xy-12y^2}{32x^2-28xy+3y^2} \div \dfrac{9x^2-24xy+16y^2}{24x^2-35xy+4y^2}$ 28._____

Name: Date:

Instructor: Section:

Perform the indicated operations. Express answers in lowest terms.

29. $\left(\dfrac{20k^2 - 9k - 18}{k^2 + 8k} \div \dfrac{5k - 6}{k^3 + 4k^2 - 32k} \right) \cdot \dfrac{k^2 - 8k + 12}{4k^2 - 5k - 6}$

29._____

30. $\left(\dfrac{12y^2 - 11y - 15}{y^2 + 7y} \div \dfrac{3y - 5}{y^2 + 3y^2 - 28y} \right) \cdot \dfrac{y^2 - 7y + 12}{4y^2 - 13y - 12}$

30._____

Given f(x) and g(x), find f(x) · g(x) and f(x) ÷ g(x). Express answers in lowest terms.

31. $f(x) = \dfrac{x - 10}{x^2 + 8x}$ and $g(x) = \dfrac{9x}{x + 8}$

31._____

_____

Objective 5   **Solve applied problems involving multiplication and division of rational expressions.**

Solve.

32. The surface area of a rectangular box is given by the expression 2*lw* + 2*lh* + 2*wh*, where *l* is the length of the box, *w* is the width of the box, and *h* is the height of the box. The expression for the volume of this box is given by *lwh*. Write the ratio of the surface area of the box to its volume in simplest form.

32._____

144

Chapter 6   RATIONAL EXPRESSIONS AND EQUATIONS

6.2 Addition and Subtraction of Rational Expressions

Learning Objectives
1 Add and subtract rational expressions with the same denominator.
2 Find the least common denominator (LCD) of two or more rational expressions.
3 Add and subtract rational expressions with different denominators.
4 Solve applied problems involving the addition or subtraction of rational expressions.

Key Terms
Use the most appropriate term or phrase from the given list to complete each statement in exercises 1-3.

    numerators   **denominators**   **simplest form**   **expanded form**

1. To find the LCD of two rational expressions, factor their _____.

2. To add two rational expressions with the same denominator, add the
    _____ and keep the _____.

3. If there are no common factors in the numerator and denominator of a rational
    expression, the expression is said to be in _____.

Objective 1 Add and subtract rational expressions with the same denominator.

Perform the indicated operation. Simplify if possible.

1. $\dfrac{7x+9}{x-5} - \dfrac{x-5}{x-5}$

                                           1._____

2. $\dfrac{4x+1}{3x^2} + \dfrac{2x-1}{3x^2}$

                                           2._____

3. $\dfrac{6c+d}{c^2d} - \dfrac{5c-d}{c^2d}$

                                           3._____

4. $\dfrac{3x-4}{x+1}+\dfrac{8x-5}{x+1}$

4._____

5. $\dfrac{6a-5n}{n^2-a^2}-\dfrac{5a-6n}{n^2-a^2}$

5._____

6. $\dfrac{6x-7y}{y^2-x^2}-\dfrac{5x-8y}{y^2-x^2}$

6._____

Objective 2 Find the least common denominator (LCD) of two or more rational expressions.

Find the LCD of each group of rational expressions. Then write each expression in terms of the LCD.

7. $\dfrac{11}{13a^3}$ and $\dfrac{4}{5a^4}$

7._____

8. $\dfrac{3}{5c^2}$ and $\dfrac{2}{3c^3}$

8._____

9. $\dfrac{10}{3x+3}$ and $\dfrac{-1}{5x+5}$

9._____

10. $\dfrac{3}{5x+15}$ and $\dfrac{-6}{7x+21}$

10._____

146

11. $\dfrac{m+8}{m^2+14m+49}$ and $\dfrac{m-7}{m^2+15m+56}$

12. $\dfrac{6}{p^2+9p+20}$, $\dfrac{5}{p^2+2p-8}$, and $\dfrac{9}{p^2+3p-10}$

Objective 3 Add and subtract rational expressions with different denominators.

Perform the indicated operations. Simplify, if possible.

13. $\dfrac{7}{9z^3y}-\dfrac{1}{21z^2y}$

14. $\dfrac{2}{p}-6p$

15. $\dfrac{n}{n+3}+\dfrac{4}{n+7}$

16. $\dfrac{5y+6}{y-8}+\dfrac{2y}{8-y}$

17. $\dfrac{10t-1}{t^2-9}+\dfrac{4}{3-t}$

17._____

18. $\dfrac{w}{w^2+11w+30}-\dfrac{5}{w^2+9w+20}$

18._____

19. $\dfrac{y}{y^2+5y+4}+\dfrac{y-3}{y^2+7y+12}$

19._____

20. $\dfrac{5}{v+7}+\dfrac{v+6}{v^2-49}+\dfrac{6}{v-7}$

20._____

Objective 4 Solve applied problems involving the addition or subtraction of rational expressions.

Solve.

21. Susan paid $175 for a lifetime membership to the zoo, so that she could gain admittance to the zoo for $1.95 per visit. For x visits, her cost per visit is $\dfrac{175}{x}+1.95$ dollars. Write this expression as a single rational expression.

21._____

Chapter 6   RATIONAL EXPRESSIONS AND EQUATIONS

6.3 Complex Rational Expressions

Learning Objectives
1   Simplify a complex rational expression.
2   Solve applied problems involving complex rational expressions.

Key Terms
Use the most appropriate term or phrase from the given list to complete each statement in exercises 1-2.

reciprocal   LCD method   algebraic   are factored

rational   division method   have no common factors other than 1

1.   Rewrite a complex fraction as the numerator divided by the denominator, then take the
_____ of the divisor and change the operation to multiplication.

2.   To simplify a complex rational expression, find the LCD of all
_____ expressions within the complex rational expression.

Objective 1 Simplify a complex rational expression.

Simplify.

1. $\dfrac{\frac{27}{3}}{8n}$

1._____

2. $\dfrac{\frac{8}{x}}{10x^7}$

2._____

3. $\dfrac{\frac{a+4}{4a^2}}{\frac{a+4}{8a}}$

3._____

4. $\dfrac{\dfrac{s^2-49}{8s^3}}{\dfrac{7-s}{5s}}$

5. $\dfrac{\dfrac{4}{x}}{9-\dfrac{1}{x}}$

6. $\dfrac{\dfrac{3}{x}}{6-\dfrac{1}{x}}$

7. $\dfrac{9-\dfrac{1}{s^2}}{3-\dfrac{1}{s}}$

8. $\dfrac{4-\dfrac{1}{v^2}}{2-\dfrac{1}{v}}$

8. _____

9. $\dfrac{\dfrac{2p-35}{p}+p}{p+7}$

9. _____

10. $\dfrac{\dfrac{1}{a}+\dfrac{1}{c}}{\dfrac{1}{a^2}-\dfrac{1}{c^2}}$

10. _____

11. $\dfrac{\dfrac{2}{7a^4}-\dfrac{1}{14a}}{\dfrac{5}{11a^2}+\dfrac{2}{33a}}$

11. _____

12. $\dfrac{\dfrac{7}{2z^4} - \dfrac{1}{4z}}{\dfrac{3}{5z^2} + \dfrac{7}{15z}}$

12._____

13. $\dfrac{1 + \dfrac{2}{y} - \dfrac{15}{y^2}}{\dfrac{1}{y} + \dfrac{5}{y^2}}$

13._____

14. $\dfrac{\dfrac{7}{x+3} + \dfrac{3}{x-9}}{\dfrac{8}{x-9} - \dfrac{7}{x+3}}$

14._____

15. $\dfrac{\dfrac{8}{z+7} + \dfrac{3}{z-9}}{\dfrac{5}{z-9} - \dfrac{6}{z+7}}$

15._____

16. $\dfrac{\dfrac{x^2-2x-24}{x^2-5x-36}}{\dfrac{x^2+6x+5}{x^2-4x-45}}$

16. _____

17. $\dfrac{\dfrac{z^2-2z-35}{z^2-z-30}}{\dfrac{z^2+5z+4}{z^2-2z-24}}$

17. _____

18. $\dfrac{x^{-4}+y^{-5}}{x^{-3}+y^{-4}}$

18. _____

19. $\dfrac{m^{-5}+n^{-3}}{m^{-4}+n^{-2}}$

19. _____

Objective 2 Solve applied problems involving complex rational expressions.

Solve.

20. Melissa can paint a house in m hours. Avra can paint a house in n hours. Working together, the time in hours it takes them to paint two houses (assume the houses are equally sized) is given by the complex rational expression $\dfrac{2}{\dfrac{1}{m}+\dfrac{1}{n}}$. Simplify this expression.

20._____

Chapter 6   RATIONAL EXPRESSIONS AND EQUATIONS

6.4 Solving Rational Equations

Learning Objectives
1   Solve an equation involving rational expressions.
2   Solve applied problems involving rational equations.

Key Terms
Use the most appropriate term from the given list to complete each statement in exercises 1-3.

GCF	**rate**	**equal**	**division**	**proportion**
LCD	**are**	**are not**	**cross-product**	

1.   Expressions contain no _____ sign, whereas equations do.

2.   Extraneous solutions _____ solutions of the original equation.

3.   To solve a rational equation, multiply each side of the equation by the
     _____.

Objective 1 Solve an equation involving rational expressions.

Solve and check.

1. $\dfrac{y}{21} - \dfrac{y}{35} = \dfrac{1}{21}$ 1._____

2. $\dfrac{k}{3} - \dfrac{k-2}{6} = \dfrac{5}{6}$ 2._____

3. $\dfrac{y+5}{y} - \dfrac{4}{3} = 0$

3._____

4. $\dfrac{2}{x+3} = \dfrac{4}{x}$

4._____

5. $\dfrac{y-5}{y-9} = \dfrac{4}{y-9}$

5._____

6. $t + \dfrac{9}{t} = -10$

6._____

7. $\dfrac{z}{4} = \dfrac{4}{z}$

7._____

8. $9 + \dfrac{8}{p} = \dfrac{20}{p^2}$

8._____

9. $\dfrac{9}{z+2} = \dfrac{8}{z-2}$ **9.** _____

10. $\dfrac{2}{z+9} = \dfrac{1}{z-5}$ **10.** _____

11. $\dfrac{z}{z-2} + \dfrac{z}{z^2-4} = \dfrac{z+3}{z+2}$ **11.** _____

12. $-\dfrac{0}{y^2-64} = -\dfrac{y+8}{y^2+8y}$ **12.** _____

Solve each equation for the indicated variable.

13. $\dfrac{B}{P} = T$ for P **13.** _____

14. $\dfrac{1}{A} = \dfrac{1}{w} + \dfrac{1}{x}$ for x **14.** _____

Name: Date:
Instructor: Section:

Solve.

15. If $f(x) = \dfrac{x-2}{x+4}$, find all values of x for which   15._____

   $f(x) = \dfrac{3}{5}$.

16. If $f(x) = \dfrac{-3x}{x+3}$ and $g(x) = \dfrac{9}{x-13}$, find all values   16._____

   of x for which $f(x) = g(x)$.

Objective 2 Solve applied problems involving rational equations.

Solve.

17. To determine the number of deer in a game   17._____
    preserve, a conservationist catches 349 deer, tags
    them and lets them loose. Later, 537 deer are
    caught; 179 of them are tagged. How many deer are
    in the preserve?

18. A barge moves 11 km/hr in still water. It travels 27   18._____
    km upriver and 27 km downriver in a total time of
    165 hr. What is the speed of the current?

158

Chapter 6   RATIONAL EXPRESSIONS AND EQUATIONS

6.5 Variation

Learning Objectives

1   Write and solve equations expressing direct variation, inverse variation, joint variation, and combined variation.

2   Solve applied problems involving variation.

Key Terms

Use the most appropriate term or phrase from the given list to complete each statement in exercises 1-2.

combined variation	directly	constant term	inversely proportional
joint variation	inversely	direct variation	directly proportional
inverse variation	constant of variation		

1. If $y = kx$, where k is a positive constant, then y is _____ to x.

2. If $y = \dfrac{k}{x}$, where k is a positive constant, the y varies _____ as x.

Objective 1   Write and solve equations expressing direct variation, inverse variation, joint variation, and combined variation.

For each pair of variables given, indicate whether the variation between the variables is direct or inverse variation.

1. The area of a parallelogram and the length of its base

     1._____

2. The length of a car in inches and the length of the car in feet

     2._____

Use the given information to find the constant of variation and the variation equation.

3. y varies directly as x; $y = 16$ when $x = 2$

     3._____

4. y varies directly as x; $y = 36$ when $x = 9$ **4.**_____

5. y varies directly as x; $y = 6$ when $x = 31$ **5.**_____

6. y varies directly as x; $y = 5$ when $x = 38$ **6.**_____

7. y varies directly as x; $y = 0.6$ when $x = 0.4$ **7.**_____

8. y varies directly as x; $y = 0.8$ when $x = 0.7$ **8.**_____

9. y varies inversely as x; $y = 7$ when $x = 18$ **9.**_____

10. y varies inversely as x; $y = 10$ when $x = 20$ **10.**_____

11. y varies inversely as x; $y = 0.3$ when $x = 0.2$ **11.**_____

12. y varies inversely as x; $y = 0.2$ when $x = 0.7$ **12.**_____

13. y varies inversely as x; $y = 5$ when $x = \dfrac{1}{5}$ **13.** _____

14. y varies inversely as x; $y = 4$ when $x = \dfrac{1}{2}$ **14.** _____

15. y varies jointly as x and z; $y = 36$ when $x = 9$ and $z = 4$ **15.** _____

16. y varies jointly as x and z; $y = 32$ when $x = 4$ and $z = 8$ **16.** _____

17. y varies jointly as w and the square of x and inversely as z; $y = 25$ when $w = 5$, $x = 5$, and $z = 15$ **17.** _____

18. y varies jointly as w and the square of x and inversely as z; $y = 4$ when $w = 1$, $x = 2$, and $z = 2$ **18.** _____

Objective 2 Solve applied problems involving variation.

Solve.

19. The number N of aluminum cans used each year
varies directly as the number of people P using
the cans. If 51 people use 13,668 cans in one year,
how many cans are used in a city which has a
population of 1,543,000?

                                               19._____

20. The wavelength W of a radio wave varies inversely
as its frequency F. A wave with a frequency of
1200 kilohertz has a length of 300 m. What is the
length of a wave with a frequency of 1000
kilohertz?

                                                 20._____

21. If the temperature is constant, the pressure of a gas
in a container varies inversely as the volume of the
container. If the pressure is 70 pounds per square
foot in a container with 7 cubic feet, what is the
pressure in a container with 4.5 cubic feet?

                                                 21._____

22. The stopping distance d of a car after the brakes
are applied varies directly as the square of the
speed r. If a car traveling 80 mph can stop in 360
ft, how many feet will it take the same car to stop
when it is traveling 20 mph?

                                                 22._____

Chapter 7 RADICAL EXPRESSIONS AND EQUATIONS

7.1 Radical Expressions and Rational Exponents

Learning Objectives
1   Evaluate Radical Expressions.
2   Write exponential expressions as radical expressions, and vice versa.
3   Solve applied problems involving radical expressions or rational exponents.

Key Terms

Use the most appropriate term or phrase from the given list to complete each statement in exercises 1-4.

is not	**square root**	**irrational**	**radicand**	**raised to the power**
rational	**multiplied by**	**cube root**	**principal**	**square**
index	**is**	**added to**	**subtracted from**	

1.   64 is an example of a perfect _____.

2.   27 _____ an example of a perfect cube.

3.   The denominator of a _____ exponent is equal to the index of the radical.

4   The power rule for exponents tells you that the exponents are _____ each other.

Objective 1   Evaluate Radical Expressions.

Evaluate, if possible.

1. $\sqrt{484}$

1._____

2. $-\sqrt{9}$

2._____

3. $\sqrt{-16}$

3._____

4. $\sqrt[3]{343}$ 4._____

5. $\sqrt[4]{10,000}$ 5._____

6. $\sqrt{\dfrac{25}{49}}$ 6._____

7. $\sqrt[3]{-\dfrac{27}{8}}$ 7._____

Use a calculator to approximate the root to the nearest thousandth.

8. $\sqrt{13}$ 8._____

9. $\sqrt{2}$ 9._____

10. $\sqrt[3]{7}$ 10._____

11. $\sqrt[3]{19}$ 11._____

Objective 2 Write exponential expressions as radical expressions, and vice versa.

Simplify. Assume that all variables represent positive real numbers.

12. $\sqrt{x^6}$ 12._____

13. $\sqrt{25x^6}$ 13._____

14. $6\sqrt{x^{16}y^{34}}$

14._____

15. $8\sqrt{x^8 y^{18}}$

15._____

16. $\sqrt[3]{-216x^6}$

16._____

17. $\sqrt[3]{-125x^9}$

17._____

18. $\sqrt[3]{27y^6}$

18._____

19. $\sqrt[4]{625y^8}$

19._____

Write using radical notation. Simplify if possible.

20. $25^{1/2}$

20._____

21. $-81^{1/4}$

21._____

22. $(-125)^{1/3}$

22._____

23. $216^{2/3}$

23._____

24. $-1000^{-5/3}$

24._____

25. $9 \cdot 9^{1/2}$

25._____

26. $\dfrac{5^{9/4}}{5^{7/4}}$

26._____

27. $q^{7/8} q^{-5/8}$

27._____

28. $\left(3x^5\right)^{-1/3}$

28._____

29. $\left(8x^{18} y^9\right)^{2/3}$

29._____

Objective 3 Solve applied problems involving radical expressions or rational exponents.

Solve.

30. After an accident, police can estimate the speed that a car was traveling by measuring its skid marks. The formula $r = 2\sqrt{5L}$ can be used, where r is the speed, in miles per hour, and L is the length of the skid marks, in feet. Estimate the speed of a car that left skid marks
    a. 170 ft long.
    b. 50 ft long.
    c. 40 ft long.

30a._____

b._____

c._____

Name: Date:
Instructor: Section:

Chapter 7 RADICAL EXPRESSIONS AND EQUATIONS

7.2 Simplifying Radical Expressions

Learning Objectives
1 Simplify radical expressions.
2 Solve applied problems involving radical expressions or rational exponents.

Key Terms
Use the most appropriate term or phrase from the given list to complete each statement in exercises 1-2.

| multiply | is | is not | subtract | add | divide |

1. To multiply radicals with the same index, _____ the radicands.

2. The expression $\sqrt[n]{a}$ _____ simplified if the radicand has a factor that is a perfect nth power.

Objective 1 Simplify radical expressions.

Use rational exponents to simplify the expression, if possible. Then write the answer in radical form. Assume that all variables represent positive real numbers.

1. $\sqrt[3]{8}$ 1._____

2. $\sqrt[8]{y^6}$ 2._____

3. $\sqrt[4]{\sqrt[5]{x}}$ 3._____

4. $\sqrt[6]{x^{24}y^3}$ 4._____

5. $\dfrac{\sqrt[5]{y^2}}{\sqrt[6]{y^2}}$ 5._____

6. $\sqrt[5]{n}\cdot\sqrt[4]{n}$ 6._____

Multiply.

7. $\sqrt{3} \cdot \sqrt{10}$

7._____

8. $\sqrt{3y} \cdot \sqrt{2z}$

8._____

9. $\sqrt[3]{11w} \cdot \sqrt[3]{10z}$

9._____

10. $\sqrt[3]{2r} \cdot \sqrt[3]{5t}$

10._____

Simplify. Assume that all variables represent positive real numbers.

11. $\sqrt{63}$

11._____

12. $\sqrt[3]{24}$

12._____

13. $-\sqrt[4]{96}$

13._____

14. $\sqrt{48x}$

14._____

15. $\sqrt{448x^6 y^{15}}$

15._____

16. $\sqrt[4]{324x^{16} y^{21}}$

16._____

Divide and simplify. Assume that all variables represent positive real numbers.

17. $\dfrac{\sqrt{20}}{\sqrt{5}}$

17._____

18. $\dfrac{\sqrt{30n}}{\sqrt{15}}$

18._____

19. $\dfrac{\sqrt{16p^3 q}}{\sqrt{4q}}$

19._____

20. $\dfrac{\sqrt[5]{54m^4 n^2}}{\sqrt[5]{9m^3 n^2}}$

20._____

Simplify.

21. $\sqrt{\dfrac{49}{81}}$

21._____

22. $\sqrt{\dfrac{121}{36}}$

22._____

23. $\sqrt{\dfrac{7}{x^6}}$

23._____

24. $\sqrt[3]{\dfrac{72x^7}{125}}$

24._____

Find the distance between the two points on the coordinate plane. Give answers in simplest radical form.

25. $(-10,-7)$ and $(0,-9)$

25._____

26. $(-4,-5)$ and $(-7,-10)$

26._____

Objective 2 Solve applied problems involving radical expressions or rational exponents.

Solve.

27. What does it mean to refer to a 20-in. TV set or a 25-in. TV set? Such units refer to the diagonal of the screen. A 30-in. TV set also has a width of 24 inches. What is its height?

27._____

28. A surveyor measured the distances shown in the figure. Find the distance across the lake between points R and S.

28._____

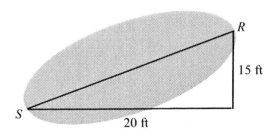

170

Chapter 7 RADICAL EXPRESSIONS AND EQUATIONS

7.3 Addition and Subtraction of Radical Expressions

Learning Objectives
1   Add and subtract radical expressions.
2   Solve applied problems involving the addition or subtraction of radical expressions.

Key Terms
Use the most appropriate term or phrase from the given list to complete each statement in exercises 1-2.

unlike	simplified	reduced	the same	like
different	radicands	prime	coefficients	powers

1. $\sqrt[3]{5}$ and $\sqrt[7]{5}$ are _____ radicals because they have _____ indices.

2.   When adding or subtracting radicals, do not combine their _____.

Objective 1 Add and subtract radical expressions.

Combine, if possible. Assume that all variables represent positive real numbers.

1. $9\sqrt{3}+3\sqrt{3}$ 1._____

2. $4\sqrt[3]{3}-2\sqrt[3]{3}$ 2._____

3. $4\sqrt{7}-2\sqrt{2}$ 3._____

4. $6\sqrt{x}-3\sqrt{x}$ 4._____

5. $9\sqrt{x} + 7\sqrt{x} - 4\sqrt{x}$ 5._____

6. $5\sqrt[3]{4} - 2\sqrt[3]{4}$ 6._____

7. $8\sqrt[4]{y} + 5\sqrt[4]{y} - 4\sqrt[4]{y}$ 7._____

8. $8y\sqrt{5x} + 9y\sqrt{5x}$ 8._____

9. $3\sqrt{x-2} + \sqrt{x-2}$ 9._____

10. $5\sqrt{a+6} - 8\sqrt{a+6}$ 10._____

Simplify, if possible.

11. $\sqrt{45} + \sqrt{125}$ 11._____

12. $\sqrt{80} - \sqrt{45}$ 12._____

13. $\sqrt{125} + \sqrt{180}$ 13._____

14. $5\sqrt{12} - \sqrt{108} + 7\sqrt{300}$

14. _____

15. $3\sqrt{200} - \sqrt{98} + 2\sqrt{18}$

15. _____

16. $3\sqrt[3]{135} - \sqrt[3]{625} + 3\sqrt[3]{5}$

16. _____

17. $8\sqrt[3]{81} - \sqrt[3]{81} + 2\sqrt[3]{3}$

17. _____

18. $9\sqrt{63b^2 a} + 2b\sqrt{252a}$

18. _____

19. $\sqrt{2a} + 7\sqrt{50a^3}$

19. _____

20. $\sqrt[3]{27xy^2} + \sqrt[3]{8x^4 y^5}$

20. _____

21. $\sqrt[4]{256xy^2} + \sqrt[4]{81x^5y^6}$ 21._____

Given f(x) and g(x), find f(x) + g(x) and f(x) − g(x).

22. $f(x) = 5x\sqrt{27x}$ and $g(x) = 4\sqrt{3x^3}$ 22._____

Objective 2   Solve applied problems involving the addition or subtraction of radical expressions.

Solve.

23. Find the length of the sides of two squares whose   23._____
areas are 448 square feet and 112 square feet. Then
find the difference of their lengths.

Name: Date:
Instructor: Section:

Chapter 7 RADICAL EXPRESSIONS AND EQUATIONS

7.4 Multiplication and Division of Radical Expressions

Learning Objectives
1   Multiply radical expressions.
2   Divide radical expressions.
3   Rationalize denominators in radical expressions.
4   Solve applied problems involving the multiplication or division of radical expressions.

Key Terms
Use the most appropriate term from the given list to complete each statement in exercises 1-2.

numerator	denominator	perfect power	constant
conjugate	product	opposite	exponent

1.   When a radical expression in a denominator contains two terms, rationalize the denominator by identifying its _____.

2.   A(n) _____ can be rationalized by multiplying the numerator and denominator by a factor that makes the radicand in the denominator a perfect power.

Objective 1 Multiply radical expressions.

Multiply and simplify

1. $\sqrt{6} \cdot \sqrt{15}$ 1._____

2. $\sqrt[4]{28a^2} \cdot \sqrt[4]{4a^5}$ 2._____

3. $\sqrt{6}\left(\sqrt{24} - 3\right)$ 3._____

4. $\sqrt{3}\left(9\sqrt{7} - 8\sqrt{4}\right)$ 4._____

5. $12\sqrt{15}\left(\sqrt{17}+3\sqrt{15}\right)$

5._____

6. $\left(\sqrt{6}-9\right)\left(\sqrt{6}-2\right)$

6._____

7. $\left(4+2\sqrt{3}\right)\left(4-2\sqrt{3}\right)$

7._____

8. $\left(\sqrt{3}-\sqrt{7}\right)\left(\sqrt{3}+\sqrt{7}\right)$

8._____

9. $\left(\sqrt{10x}+5\right)\left(\sqrt{10x}-5\right)$

9._____

10. $\left(\sqrt{3u}+\sqrt{v}\right)\left(\sqrt{3u}-\sqrt{v}\right)$

10._____

Objective 2 Divide radical expressions.

Divide and simplify. Assume that all variables represent positive real numbers.

11. $\dfrac{\sqrt{80n^{15}}}{\sqrt{5n}}$

11._____

12. $\dfrac{\sqrt{28n^{9}}}{\sqrt{7n}}$

12._____

13. $\sqrt{\dfrac{y^8}{49}}$

13._____

14. $\sqrt{\dfrac{x^4}{25}}$

14._____

15. $\sqrt{\dfrac{20x^7}{y^5}}$

15._____

Objective 3 Rationalize denominators in radical expressions.

Simplify by rationalizing the denominator. Assume that all variables represent positive real numbers.

16. $\dfrac{2}{\sqrt{5}}$

16._____

17. $\dfrac{\sqrt{2x}}{\sqrt{27}}$

17._____

18. $\dfrac{\sqrt[3]{5a}}{\sqrt[3]{2c}}$

18._____

19. $\sqrt{\dfrac{6}{x}}$

19._____

20. $\sqrt{\dfrac{16y^5}{x^8}}$ 20._____

21. $\dfrac{1}{5+\sqrt{7}}$ 21._____

22. $\dfrac{5}{\sqrt{5}-\sqrt{10}}$ 22._____

23. $\dfrac{\sqrt{a}-\sqrt{b}}{\sqrt{a}+\sqrt{b}}$ 23._____

24. $\dfrac{\sqrt{7}-9}{\sqrt{5}}$ 24._____

Objective 4 Solve applied problems involving the multiplication or division of radical expressions.

Solve.

25. Find the area of a rectangle whose length is $\sqrt{4}$ in.   25._____
and width is $2\sqrt{4}$ in.

Chapter 7 RADICAL EXPRESSIONS AND EQUATIONS

7.5 Solving Radical Equations

Learning Objectives
1   Solve radical equations.
2   Solve applied problems involving radical equations.

Objective 1   Solve radical equations.

Solve.

1. $\sqrt{x} = 5$

1._____

2. $\sqrt{x+1} = 11$

2._____

3. $\sqrt{5y+6} = 7$

3._____

4. $\sqrt[3]{x+9} = 5$

4.   _____

5. $\sqrt{x} + 8 = 7$

5._____

6. $\sqrt{x} - 6 = 8$

6._____

7. $\sqrt{5x} + 8 = 15$ 7._____

8. $\sqrt{y+3} - 4 = 6$ 8._____

9. $\sqrt{4y+4} - \sqrt{3y+7} = 0$ 9._____

10. $\sqrt{9m+9} - \sqrt{8m+10} = 0$ 10._____

11. $\sqrt{2y+3} = \sqrt{2y-2}$ 11._____

12. $\sqrt[7]{5x-7} = \sqrt[7]{8x+35}$ 12._____

13. $\sqrt[5]{4x-10} = \sqrt[5]{7x+20}$ 13._____

14. $\sqrt[3]{a^2-5}+\sqrt[3]{1-3a}=0$

14._____

15. $\sqrt{9x+67}=x+5$

15._____

16. $\sqrt{7x+46}-4=x$

16._____

17. $3+\sqrt{z-1}=-\sqrt{z+14}$

17._____

18. $\sqrt{3x+24}-2=x$

18._____

19. $3+\sqrt{z-3}=\sqrt{z+12}$

19._____

20. $\sqrt{x+3}+\sqrt{3x+13}=2$

20._____

Objective 2 Solve applied problems involving radical equations.

Solve.

21. The formula $r = 2\sqrt{5L}$ can be used to
    approximate the speed, r in miles per hour, of a car
    that has left skid marks of length L, in feet. How
    far will a car skid at 60 mph?

    21._____

22. A 26-ft ladder is placed against a vertical wall of a
    building, with the bottom of the ladder standing on
    level ground 24 ft from the base of the building.
    How high up the wall does the ladder reach?

    22._____

Chapter 7 RADICAL EXPRESSIONS AND EQUATIONS

7.6 Complex Numbers

Learning Objectives
1   Identify imaginary and complex numbers.
2   Add and subtract complex numbers.
3   Multiply complex numbers.
4   Identify a complex conjugate.
5   Divide complex numbers.
6   Evaluate powers of i.
7   Solve applied problems involving complex numbers.

Key Terms
Use the most appropriate term from the given list to complete each statement in exercises 1-2.

complex number	irrational number	imaginary part	power
imaginary number	complex conjugates	real part	multiple

1.   In the complex number $7 + 3i$, 7 is the real part and 3 is the _____.

2.   Every real number is a(n) _____.

Objective 1 Identify imaginary and complex numbers.

Write in terms of i.

1. $\sqrt{-64}$ 1._____

2. $\sqrt{-20}$ 2._____

3. $\sqrt{-80}$ 3._____

4. $-\sqrt{-16}$ 4._____

Objective 2 Add and subtract complex numbers.

Perform the indicated operation.

5. $(6+3i)+(9-2i)$ 5._____

6. $(9+6i)+(4-5i)$ 6._____

7. $(9+5i)-(-1-i)$ 7._____

8. $(2+3i)-(-3-i)$ 8._____

9. $16-\left(2+\sqrt{-36}\right)$ 9._____

Objective 3 Multiply complex numbers.

Multiply.

10. $\sqrt{-16}\cdot\sqrt{-4}$ 10._____

11. $5i\cdot 4i$ 11._____

12. $-6i\cdot 2i$ 12._____

13. $3i(-2+3i)$

13._____

14. $\sqrt{-4}\left(6+\sqrt{-16}\right)$

14._____

15. $(6+8i)(3-9i)$

15._____

16. $(4+2i)(4-2i)$

16._____

17. $(3+5i)^2$

17._____

Objective 4 Identify a complex conjugate.

Find the complex conjugate of the complex number. Then find the product of each complex number and its complex conjugate.

18. $2+4i$

18._____

19. $7+2i$

19._____

Objective 5 Divide complex numbers.

Divide.

20. $\dfrac{4}{5+i}$

20._____

21. $\dfrac{7}{9i}$

21._____

22. $\dfrac{5-10i}{50i}$

22._____

23. $\dfrac{2+4i}{5+9i}$

23._____

Objective 6 Evaluate powers of *i*.

Evaluate.

24. i^{42}

24._____

25. i^{21}

25._____

26. i^{35}

26._____

Objective 7 Solve applied problems involving complex numbers.

Solve.

27. The total impedance in an AC circuit connected in series
is the sum of the individual impedances. If the
impedance in one part of the circuit is $(13-2i)$ ohms
and in another part is $(9+4i)$ ohms, what is the total
impedance in the circuit?

27._____

Chapter 8 QUADRATIC EQUATIONS, FUNCTIONS, AND INEQUALITIES

8.1 Solving Quadratic Equations by Completing the Square

Learning Objectives
1   Solve quadratic equations using the square root property.
2   Solve quadratic equations by completing the square.
3   Solve applied problems involving quadratic equations.

Objective 1   Solve quadratic equations using the square root property.

Solve by using the square root property of equality.

1. $x^2 = 49$ 1._____

2. $t^2 = 75$ 2._____

3. $x^2 + 36 = 0$ 3._____

4. $3x^2 - 15 = 0$ 4._____

5. $2 - 4y^2 = 32$ 5._____

6. $(x - 3)^2 = 45$ 6._____

7. $(5k + 1)^2 = 45$ 7._____

8. $(2k + 1)^2 = 20$ 8._____

9. $(3k - 1)^2 + 9 - 1$ 9._____

10. $3 - 2x = \dfrac{22}{3 - 2x}$

Solve the formula for the indicated variable.

11. $y = \dfrac{1}{3}\pi r^2 x$ for r

11._____

12. $F = \dfrac{Jm_1 m_2}{r^2}$ for r

12._____

Objective 2 Solve quadratic equations by completing the square.

Find the number that should be added to the expression to make it a perfect square trinomial.

13. $y^2 + 2y$

13._____

14. $z^2 + 6z$

14._____

15. $k^2 - 5k$

15._____

16. $x^2 - \dfrac{1}{4}x$

16._____

Solve by completing the square.

17. $x^2 - 4x = -3$

17._____

18. $x^2 - 8x = -15$

18._____

19. $x^2 + 2x - 6 = 0$

19._____

20. $x^2 + 2x - 2 = 0$

20._____

21. $3x^2 + 20x - 3 = 0$

21._____

22. $3r^2 + 4r + 11 = 0$

22._____

23. $x^2 + \dfrac{1}{2}x - 1 = 0$

23._____

24. $5r^2 - r - 4 = 0$

24._____

25. $4x^2 - x - 3 = 0$

25._____

Given f(x) and g(x), find all values of x such that f(x) = g(x).

26. $f(x) = x^2 - 8$ and $g(x) = 8x - 11$

26._____

27. $f(x)=x^2-10$ and $g(x)=6x-6$ **27.**_____

Objective 3 Solve applied problems involving quadratic equations.

Solve.

28. The formula $A=P(1+r)^2$ gives the amount A in   **28.**_____
dollars that P will grow to in 2 years at interest rate
r (where r is given as a decimal), using compound
interest. What interest rate will cause $3000 to
grow to $3213.68 in 2 years? Round answer to the
nearest tenth of a percent.

29. The formula $S=16t^2$ is used to approximate the   **29.**_____
distance S, in feet, that an object falls freely from
rest in t seconds. The height of the building is 1345
feet. How long would it take for an object to fall
from the top? Round answer to the nearest tenth of
a second.

30. Two cars leave town, one driving north and the   **30.**_____
other east. They are 52 miles apart when one of
them is 28 miles farther from town than the other.
At that time, how far were they each from the
town?

Chapter 8 QUADRATIC EQUATIONS, FUNCTIONS, AND INEQUALITIES

8.2 Solving Quadratic Equations by Using the Quadratic Formula

Learning Objectives
1. Solve quadratic equations using the quadratic formula.
2. Determine the number and type of solutions to a quadratic equation using the discriminant.
3. Solve applied problems using the quadratic formula.

Key Terms
Use the most appropriate term or phrase from the given list to complete each statement in exercises 1-2.

> one   quadratic equation   negative   discriminant   numerator
>
> no   quadratic formula   positive

1. If the discriminant of a quadratic equation is _____, the equation has two complex solutions containing i.

2. Whenever the solutions to a _____ are rational numbers, the original equation can also be solved by factoring.

Objective 1 Solve quadratic equations using the quadratic formula.

Solve.

1. $x^2 - 2x - 15 = 0$ 1._____

2. $x^2 - 5x - 50 = 0$ 2._____

3. $x^2 + 10x + 4 = 0$ 3._____

4. $x^2 - 7x - 60 = 0$ 4._____

5. $x^2 + 18x + 4 = 0$ **5.**＿＿＿＿＿＿＿＿＿

6. $x^2 - 6x + 19 = 6$ **6.**＿＿＿＿＿＿＿＿＿

7. $x^2 - 2x + 47 = 10$ **7.**＿＿＿＿＿＿＿＿＿

8. $3x^2 - 5x = 7$ **8.**＿＿＿＿＿＿＿＿＿

9. $9x^2 - 3x = 5$ **9.**＿＿＿＿＿＿＿＿＿

10. $2x^2 - x + 6 = 0$ **10.**＿＿＿＿＿＿＿＿＿

11. $x^2 - x + 9 = 0$ **11.**＿＿＿＿＿＿＿＿＿

12. $3x^2 - 8x + 22 = 6 + 16x - 6x^2$ **12.**＿＿＿＿＿＿＿＿＿

13. $6x^2 - 5x + 30 = 5 + 25x - 3x^2$ **13.**＿＿＿＿＿＿＿＿＿

14. $\dfrac{x^2}{4} - \dfrac{x}{2} = 1$ 14._____

15. $\dfrac{x^2}{14} - \dfrac{x}{7} = 1$ 15._____

16. $\dfrac{1}{2}x^2 + \dfrac{1}{10}x - 1 = 0$ 16._____

17. $\dfrac{1}{3}x^2 + \dfrac{1}{21}x - 1 = 0$ 17._____

18. $0.2x^2 - 0.4x - 0.2 = 0$ 18._____

19. $(x+3)(x-4) = -8$ 19._____

20. $3.7x^2 + 4.4x - 5.6 = 0$ 20._____
      Round answers to the nearest hundredth.

Objective 2 Determine the number and type of solutions to a quadratic equation using the discriminant.

Use the discriminant to determine the number and type of solution(s) for each equation..

21. $x^2 - 4x + 7 = 0$ 21._____

22. $9x^2 - 6x = -1$ 22._____

23. $2x^2 + 6x + 4 = 0$ 23._____

24. $x^2 - 3x + 5 = 0$ 24._____

Objective 3 Solve applied problems using the quadratic formula.

Solve.

25. A certain bakery has found that the daily demand   25._____
for bran muffins is $\dfrac{12,600}{p}$, where p is the price of
a muffin in cents. The daily supply is $4p - 100$.
Find the price at which supply and demand are
equal.

Chapter 8 QUADRATIC EQUATIONS, FUNCTIONS, AND INEQUALITIES

8.3 More on Quadratic Equations

Learning Objectives
1   Solve an equation that leads to a quadratic equation.
2   Determine a quadratic equation with given solutions.
3   Solve applied problems involving quadratic equations.

Objective 1 Solve an equation that leads to a quadratic equation.

Solve and check.

1. $10 - \dfrac{1}{t} - \dfrac{9}{t^2} = 0$

1._____

2. $6 - \dfrac{1}{n} - \dfrac{5}{n^2} = 0$

2._____

3. $x^4 - 7x^2 + 6 - 0$

3._____

4. $y^4 - 6y^2 + 8 = 0$

4._____

5. $p - 2\sqrt{p} - 8 = 0$

5._____

6. $m - 3\sqrt{m} - 10 = 0$

6._____

7. $y^{1/2} - y^{1/4} - 6 = 0$ 7._____

8. $c^{1/3} - c^{1/6} - 12 = 0$ 8._____

9. $x^{2/3} - 4x^{1/3} - 5 = 0$ 9._____

10. $z^{2/3} - 2z^{1/3} - 3 = 0$ 10._____

11. $(t + 4)^2 - 13(t + 4) + 42 = 0$ 11._____

12. $(r + 2)^2 - 11(r + 2) + 28 = 0$ 12._____

Objective 2   Determine a quadratic equation with given solutions.

Find a quadratic equation with integer coefficients that has the given solutions.

13. $x = -10$, $x = 2$ 13._____

14. $y = -6$, $y = 3$ 14._____

15. $m = 5,\ m = \dfrac{4}{5}$ **15.**_____

16. $x = 3,\ x = \dfrac{3}{4}$ **16.**_____

17. $s = -\dfrac{6}{7},\ s = \dfrac{9}{7}$ **17.**_____

18. $v = -\dfrac{1}{5},\ v - \dfrac{7}{5}$ **18.**_____

19. $x = 10$ **19.**_____

20. $m = 2$ **20.**_____

21. $t = -\sqrt{14}$, $t = \sqrt{14}$ **21.**_____

22. $k = -\sqrt{13}$, $k = \sqrt{13}$ **22.**_____

23. $x = -2i$, $x = 2i$ **23.**_____

24. $x = -5i$, $x = 5i$ **24.**_____

Objective 3   Solve applied problems involving quadratic equations.

Solve.

25. During the first part of a trip, a canoeist travels 80   **25.**_____
miles at a certain speed. The canoeist travels 35
miles on the second part of the trip at a speed 5 mph
slower. The total time for the trip is 3 hrs. What was
the speed on each part of the trip?

Chapter 8 QUADRATIC EQUATIONS, FUNCTIONS, AND INEQUALITIES

8.4 Graphing Quadratic Functions

Learning Objectives
1   Graph quadratic functions.
2   Identify the vertex, the axis of symmetry, and the intercepts of a parabola.
3   Solve applied problems relating to the graph of a quadratic function.

Key Terms
Use the most appropriate term from the given list to complete each statement in exercises 1-3.

| *x*-coordinate | minimum | upward |
| *y*-coordinate | maximum | downward |

1.   When a is positive, the parabola given by the function $f(x) = ax^2 + bx + c$ has a(n) _____ point.

2.   When a is negative, the parabola given by the function $f(x) = ax^2 + bx + c$ opens _____.

3.   For a parabola given by the function $f(x) = ax^2 + bx + c$, the _____ of the vertex is $f\left(-\dfrac{b}{2a}\right)$.

Objective 1 Graph quadratic functions.

Complete the table Then plot the points and sketch the graph of the parabola.

1.

x	$y = f(x) = -3x^2$	(x, y)
-2		
-1		
0		
1		
2		

1.

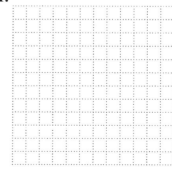

2.

x	$y = f(x) = 2x^2$	(x, y)
-2		
-1		
0		
1		
2		

2.

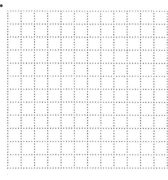

3.

x	$y = f(x) = \dfrac{1}{4}x^2$	(x, y)
-4		
-2		
0		
2		
4		

3.

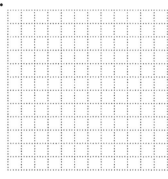

4.

x	$y = f(x) = 3 - x^2$	(x, y)
-2		
-1		
0		
1		
2		

4.

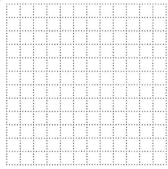

Objective 2 Identify the vertex, the axis of symmetry, and the intercepts of a parabola.

For each function, find the vertex, the axis of symmetry, and the x- and y-intercepts. Then sketch the graph.

5. $f(x) = -x^2 - 2x + 3$

5._____

_____

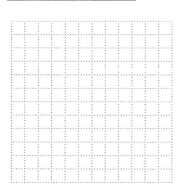

6. $f(x) = x^2 + 4x - 5$

6._____

_____

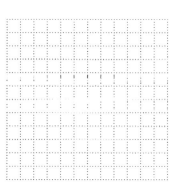

7. $f(x) = x^2 - 16$

7._____

_____

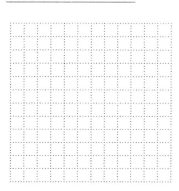

8. $f(x) = 9 - x^2$

8. _____

_____

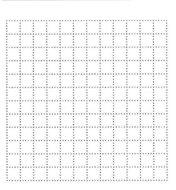

9. $f(x) = (x+8)^2$

9. _____

_____

10. $f(x) = (x-4)^2$

10. _____

_____

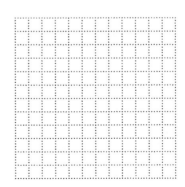

11. $f(x) = -x^2 + 4x + 5$

11._____

_____

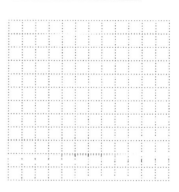

12. $f(x) = x^2 + 2x + 1$

12._____

_____

Graph each function. Then determine the domain and range.

13. $f(x) = 3x^2 - 18x + 5$

13._____

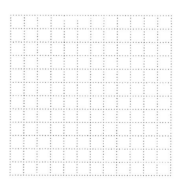

14. $g(x) = 2x^2 - 12x + 15$

14._____

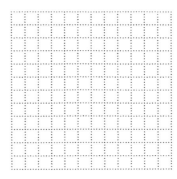

15. $f(x) = x^2 - 2x - 5$

15._____

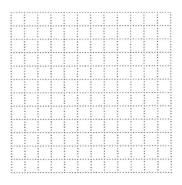

Objective 3 Solve applied problems relating to the graph of a quadratic function.

Solve.

16. A toy rocket is launched from the top of a building 360 feet tall at an initial velocity of 112 feet per second. The height of the rocket t seconds after launch is given by the equation $s(t) = -16t^2 + 112t + 360$. When does the rocket reach its maximum height? What is the maximum height?

16._____

17. A carpenter is building a rectangular room with a fixed perimeter of 260 feet. What dimensions would yield the maximum area? What is the maximum area?

17._____

Chapter 8 QUADRATIC EQUATIONS, FUNCTIONS, AND INEQUALITIES

8.5 Solving Quadratic and Rational Inequalities

Learning Objectives
1 Solve quadratic and rational inequalities.
2 Solve applied problems involving quadratic or rational inequalities.

Key Terms
Use the most appropriate term from the given list to complete each statement in exercises 1-2.

boundary points	quadratic	denominator	numerator
test values	equation	ratio	rational inequality

1.   The key to solving a quadratic inequality is to solve the related _____.

2.   _____ separate the number line into intervals, from which test values are selected.

Objective 1 Solve quadratic and rational inequalities.

Solve. Write solutions in interval notation Then graph the solutions.

1. $x^2 < 9$

1._____

← $|$ $|$ $|$ $|$ $|$ $|$ $|$ $|$ $|$ $|$ →

2. $x^2 \leq 36$

2._____

← $|$ $|$ $|$ $|$ $|$ $|$ $|$ $|$ $|$ $|$ $|$ →

3. $x^2 - 12x + 36 \geq 0$

3._____

← $|$ $|$ $|$ $|$ $|$ $|$ $|$ $|$ $|$ $|$ $|$ →

4. $x^2 - 7x - 18 \geq 0$ 4._____

$\longleftarrow\!+\!+\!+\!+\!+\!+\!+\!+\!+\!+\!+\!+\!\longrightarrow$

5. $x^2 - 5x - 24 \geq 0$ 5._____

$\longleftarrow\!+\!+\!+\!+\!+\!+\!+\!+\!+\!+\!+\!+\!\longrightarrow$

6. $x^2 - 16x + 64 \geq 0$ 6._____

$\longleftarrow\!+\!+\!+\!+\!+\!+\!+\!+\!+\!+\!+\!+\!\longrightarrow$

7. $x^2 - 10x + 21 < 0$ 7._____

$\longleftarrow\!+\!+\!+\!+\!+\!+\!+\!+\!+\!+\!+\!+\!\longrightarrow$

8. $x^2 - 16x + 60 < 0$ 8._____

$\longleftarrow\!+\!+\!+\!+\!+\!+\!+\!+\!+\!+\!+\!+\!\longrightarrow$

9. $14x^2 + 5x - 1 \geq 0$ 9._____

$\longleftarrow\!+\!+\!+\!+\!+\!+\!+\!+\!+\!+\!+\!+\!\longrightarrow$

10. $8x^2 + 2x - 1 \geq 0$ 10._____

$\longleftarrow\!+\!+\!+\!+\!+\!+\!+\!+\!+\!+\!+\!+\!\longrightarrow$

11. $15x^2 + 2x \geq 1$

11._____

←+++++++++++++→

12. $12x^2 + 4x \geq 1$

12._____

←+++++++++++++→

13. $\dfrac{1}{x-7} < 0$

13._____

←+++++++++++++→

14. $\dfrac{1}{x-2} > 0$

14._____

←+++++++++++++→

15. $\dfrac{x+2}{x+6} \geq 0$

15._____

←+++++++++++++→

16. $\dfrac{x+1}{x+4} \geq 0$

16._____

←+++++++++++++→

17. $\dfrac{2x+3}{x-6} \le 0$

←+++++++++++++→

18. $\dfrac{3x+4}{x-4} \le 0$

←+++++++++++++→

Objective 2 Solve applied problems involving quadratic or rational inequalities.

Solve.

19. A toy rocket is shot upward from the ground with an initial velocity of 120 ft per sec. The height, h, of the rocket above the ground t sec after it is launched is given by $h(t) = -16t^2 + 120t$. For what values of t is the rocket at least 144 ft above the ground?

19._____

20. The base of a storage box has a perimeter of 120 in. For what range of lengths will the area of the base be at least 275 in.^{2}?

20._____

Chapter 9 EXPONENTIAL AND LOGARITHMIC FUNCTIONS

9.1 The Algebra of Functions and Inverse Functions

Learning Objectives
1   Add, subtract, multiply, and divide functions.
2   Find the composition of two functions.
3   Determine if a function is one-to-one.
4   Find the inverse of a function.
5   Graph a function and its inverse.
6   Solve applied problems involving functions.

Key Terms

Use the most appropriate term or phrase from the given list to complete each statement in exercises 1-4.

composition	difference	linear	$g(x)$	g	first	second
horizontal	vertical	inverse	$f(x)$	f	sum	product

1. $(f \circ g)(x)$ is defined for those values of x in the domain of _____ where

    _____ is also in the domain of _____.

2. A function is one-to-one if no two ordered pairs have the same_____
    coordinates.

3. To test whether a function is one-to-one, examine its graph and use the
    _____ line test.

4. The _____ of a one-to-one function is formed by interchanging the
    coordinates of the ordered pairs that define the original function.

Objective 1   Add, subtract, multiply, and divide functions.

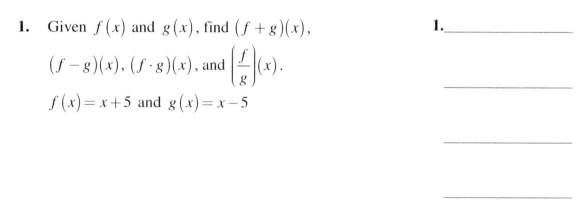

1. Given $f(x)$ and $g(x)$, find $(f+g)(x)$,   1._____

    $(f-g)(x)$, $(f \cdot g)(x)$, and $\left(\dfrac{f}{g}\right)(x)$.

    $f(x) = x+5$ and $g(x) = x-5$ _____

                                                      _____

                                                      _____

2. Given $f(x)$ and $g(x)$, find $(f+g)(x)$,

$(f-g)(x)$, $(f \cdot g)(x)$, and $\left(\dfrac{f}{g}\right)(x)$.

$f(x) = 2x^2 - 4x + 5$ and $g(x) = x^3$

2._____

_____

_____

_____

3. Given $f(x)$ and $g(x)$, find $\left(\dfrac{f}{g}\right)(x)$.

$f(x) = \dfrac{4}{x+2}$ and $g(x) = \dfrac{1}{8-x}$

3._____

4. Given $f(x) = x - 3$ and $g(x) = x^2 - x$, find $(f+g)(-2)$.

4._____

5. Given $f(x) = x^2 - 1$ and $g(x) = 3 - x$, find $(f-g)(-7)$.

5._____

6. Given $f(x) = x^2 - 5$ and $g(x) = 2x + 1$, find $(f \cdot g)\left(-\dfrac{1}{2}\right)$.

6._____

7. Given $f(x) = x^2 - 17$ and $g(x) = 1 - x$, find $\left(\dfrac{f}{g}\right)(0)$.

7._____

Objective 2   Find the composition of two functions.

Given f(x) and g(x), find (f ∘ g)(x) and (g ∘ f)(x).

8. $f(x) = 7x - 5$ and $g(x) = 8 - 3x$ **8.**_____

                                                   _____

9. $f(x) = 2x^2 + 5$ and $g(x) = 3x - 4$ **9.**_____

                                                   _____

10. $f(x) = 6x^2 - 5$ and $g(x) = \dfrac{8}{x}$ **10.**_____

Objective 3   Determine if a function is one-to-one.

Determine whether the function whose graph is shown is a one-to-one function.

11.   **11.**_____

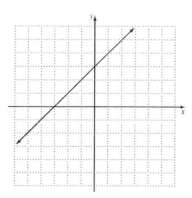

12.

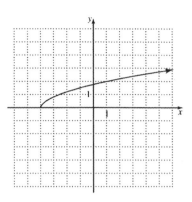

12._____

13.

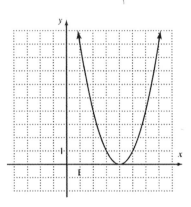

13._____

14.

14._____

Objective 4 Find the inverse of a function.

15. Find the inverse of the relation
$\{(-9,5),(1,-6),(8,9),(5,-5)\}$.

15._____

Find the inverse of each of the following one-to-one functions.

16. $f(x) = x + 9$ 16._____

17. $g(x) = x - 3$ 17._____

18. $g(x) = 3x + 8$ 18._____

19. $f(x) = x^3 - 1$ 19._____

20. $h(x) = \dfrac{9}{x + 8}$ 20._____

21. $g(x) = \dfrac{x - 3}{x + 7}$ 21._____

22. $f(x) = \sqrt[4]{x - 5}$ 22._____

Determine whether the two given functions are inverses of each other.

23. $f(x) = 5x + 6$ and $g(x) = \dfrac{x - 6}{5}$ 23._____

24. $f(x) = \dfrac{1}{x+2}$ and $g(x) = \dfrac{1}{x-2}$

24._____

Objective 5 Graph a function and its inverse.

Given the graph of a one-to-one function, sketch the graph of its inverse.

25.

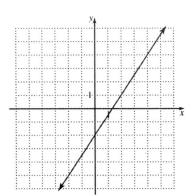

25.

26.

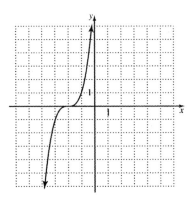

26.

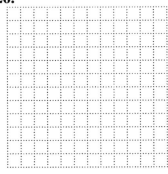

Objective 6 Solve applied problems involving functions.

Solve.

27. An organization determines that the cost per person of chartering a bus is given by the formula $C(x) = \dfrac{400+5x}{x}$, where x is the number of people in the group and $C(x)$ is in dollars. What does $C^{-1}(18)$ represent?

27._____

Chapter 9 EXPONENTIAL AND LOGARITHMIC FUNCTIONS

9.2 Exponential Functions

Learning Objectives
1 Evaluate exponential functions.
2 Graph exponential functions.
3 Solve exponential equations of the form $b^x = b^n$.
4 Solve applied problems involving exponential functions.

Key Terms
Use the most appropriate term or phrase from the given list to complete each statement in exercises 1-2.

imaginary	**decreasing**	**crosses**	**approaches**
irrational	**positive**	**real**	**increasing**

1. The natural exponential function is defined for all _____ numbers.

2. The exponential function $f(x) = b^x$ is _____ if the base is greater than 1.

Objective 1 Evaluate exponential functions.

Determine whether each function is a polynomial function, a rational function, a radical function, or an exponential function.

1. $f(x) = 2x^{1/4} - 1$ 1._____

2. $f(x) = 3 \cdot 2^x - 2$ 2._____

3. $f(x) = 3x^5 - 5$ 3._____

4. $f(x) = x^{1/3}$ 4._____

Evaluate each function for the given values.

5. $f(x) = 4^x$; $x = 3$ 5._____

6. $g(x) = \left(\dfrac{1}{7}\right)^x$; $x = 4$

6._____

7. $g(x) = 5^{1-x}$; $x = -2$

7._____

8. $f(x) = 5^{x+2}$; $x = 3$

8._____

Evaluate each function for the given values. Round to the nearest thousandth.

9. $f(x) = -e^x$; $x = 2$

9._____

10. $g(x) = e^{-3x}$; $x = -\dfrac{1}{4}$

10._____

11. $h(x) = e^{-x}$; $x = 4$

11._____

12. $f(x) = e^{3x+1}$; $x = -\dfrac{1}{3}$

12._____

Objective 2 Graph exponential functions.

Graph each function.

13. $y = 6^x$

13.

14. $y = \left(\dfrac{1}{7}\right)^x$

14.

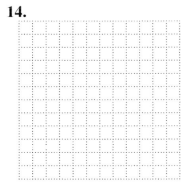

15. $y = -4^x$

15.

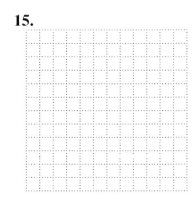

16. $y = 4^{x+4}$

16.

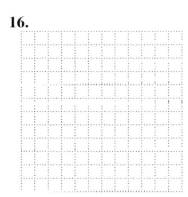

17. $y = \left(\dfrac{1}{3}\right)^x - 3$

17.

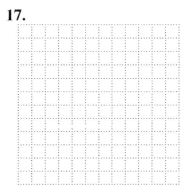

Objective 3 Solve exponential equations of the form $b^x = b^n$.

Solve.

18. $11^x = 1331$

18._____

19. $3^x = \dfrac{1}{27}$

19._____

20. $27^x = 81$

20._____

21. $2^{2x-3} = 16$

21._____

22. $256^{3x} = 64^{x+2}$

22._____

Objective 4 Solve applied problems involving exponential functions.

Solve.

23. How much money will there be in an account at the end of 4 years if $10,000 is deposited at 5%? The interest is compounded semi-annually.

23._____

24. Suppose 3000 bacteria are present at time $t = 0$. Then t minutes later, the number of bacteria present will be $N(t) = 3000(2)^{t/30}$. Find the number of bacteria present after 15 and 90 minutes.

24._____

Chapter 9 EXPONENTIAL AND LOGARITHMIC FUNCTIONS

9.3 Logarithmic Functions

Learning Objectives
1. Write equivalent exponential and logarithmic equations.
2. Evaluate logarithms.
3. Solve logarithmic equations.
4. Graph logarithmic functions.
5. Solve applied problems involving logarithms.

Key Terms

Use the most appropriate term or phrase from the given list to complete each statement in exercises 1-2.

exponential	**unknown**	**rational**	**polynomial**
logarithmic	**the base** b	**exponent**	x

1. The expression $\log_b x$ is the exponent to which _____ must be raised to get _____.

2. To solve an equation involving logarithms, write the equivalent _____equation and then solve for the _____..

Objective 1 Write equivalent exponential and logarithmic equations.

Write each logarithmic equation in its equivalent exponential form.

1. $\log_5 25 = 2$ 1._____

2. $\log_{10} \dfrac{1}{1,000,000} = -6$ 2._____

3. $\log_3 \dfrac{1}{9} = -2$ 3._____

Write each exponential equation in its equivalent logarithmic form.

4. $5^3 = 125$

4._____

5. $\left(\dfrac{1}{10}\right)^{-3} = 1000$

5._____

6. $625^{1/4} = 5$

6._____

Objective 2   Evaluate logarithms.

Evaluate.

7. $\log_4 16$

7._____

8. $\log_2 8$

8._____

9. $\log_5 1$

9._____

10. $\log_3\left(\dfrac{1}{3}\right)$

10._____

11. $\log_6\left(\dfrac{1}{1296}\right)$

11._____

12. $\log_{39} 1$

12._____

Objective 3 Solve logarithmic equations.

Solve.

13. $\log_2 x = 3$

13._____

14. $\log_2 x = -3$

14._____

15. $\log_8 x = \dfrac{1}{3}$

15._____

16. $\log_x 16 = 4$

16._____

17. $\log_x 5 = \dfrac{1}{2}$

17._____

18. $\log_x 81 = -4$

18._____

19. $x = \log_{16} 4$

19._____

20. $\log_{25} 125 = x$ **20.**_____

Objective 4   Graph logarithmic functions.

Graph each function.

21. $f(x) = \log_5 x$ **21.**

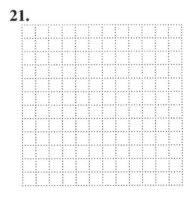

22. $f(x) = \log_{1/3} x$ **22.**

Objective 5 Solve applied problems involving logarithms.

Solve.

23. The loudness, L, in decibels (dB), is given by the formula   **23.**_____
$L = \log \dfrac{I}{I_0}$, where $I_0 = 10^{-12}$ W/m^2. The intensity, I, of a
sound is 4.1×10^{-6} W/m^2. How loud in decibels is this
sound level? Round to the nearest whole number if
needed.

Name:
Instructor:

Date:
Section:

Chapter 9 EXPONENTIAL AND LOGARITHMIC FUNCTIONS

9.4 Properties of Logarithms

Learning Objectives
1. Use the product property of logarithms.
2. Use the quotient property of logarithms.
3. Use the power property of logarithms.
4. Solve applied problems using the properties of logarithms.

Key Terms

Use the most appropriate term from the given list to complete each statement in exercises 1-2.

base   numerator   sum   denominator   product

minus   power   times   number   difference

1. The logarithm of a _____ is the sum of the logarithms of the factors.

2. The logarithm of a quotient is the logarithm of the _____ minus the logarithm of the _____.

Objective 1 Use the product property of logarithms.

Write each expression as the sum of logarithms.

1. $\log_4(64\cdot11)$ 1._____

2. $\log_5 13z$ 2._____

3. $\log_c Dx$ 3._____

4. $\log_z[u(v-1)]$ 4._____

Write each expression as a single logarithm.

5. $\log_3 8 + \log_3 5$ 5._____

6. $\log_7 9 + \log_7 8$ 6._____

7. $1 + \log_x 9$ 7._____

8. $\log_c F + \log_c x$ 8._____

Objective 2 Use the quotient property of logarithms.

Write each expression as the difference of logarithms.

9. $\log_4 \dfrac{51}{64}$ 9._____

10. $\log_6 \dfrac{x}{y}$ 10._____

11. $\log_m \dfrac{m+2}{m+3}$ 11._____

12. $\log_{10} \dfrac{10{,}000}{x}$ 12._____

Write each expression as a single logarithm.

13. $\log_4 9 - \log_4 3$ 13._____

14. $\log_7 5 - \log_7 20$ 14._____

15. $1 - \log_k 14$ 15._____

16. $\log_3 y - \log_3 (w - y)$ 16._____

Objective 3 Use the power property of logarithms.

Use the power property to rewrite each expression.

17. $\log_9 s^{10}$ 17._____

18. $\log_N r^{-10}$ 18._____

19. $\log_6 \sqrt[3]{c}$ 19._____

Evaluate.

20. $\log_r r^{14}$ 20._____

21. $11^{\log_{11} d}$ 21._____

Write each expression as a single logarithm.

22. $4\log_a 5 - 3\log_a 2$

22._____

23. $3\log_b x - 5\log_b y^4$

23._____

24. $\frac{1}{2}\left[\log_3\left(m^2 - n^2\right) - \log_3\left(m - n\right)\right]$

24._____

Write each expression as the sum or difference of logarithms.

25. $\log_c \dfrac{7x^7}{y^6}$

25._____

26. $3\log_b x + 5\log_b y + \log_b z$

26._____

Objective 4 Solve applied problems using the properties of logarithms.

Solve.

27. The formula for the pH of a solution of hydronium ions is given by the logarithmic equation $pH = -\log_{10}\left[H_3O^+\right]$, where $\left[H_3O^+\right]$ is the concentration of hydronium ions. The concentration of hydronium ions is 10^{-3}. Find the pH for the given hydronium ion $\left[H_3O^+\right]$ concentration.

27._____

Chapter 9 EXPONENTIAL AND LOGARITHMIC FUNCTIONS

9.5 Common Logarithms, Natural Logarithms, and Change of Base

Learning Objectives
1 Evaluate common logarithms.
2 Evaluate natural logarithms.
3 Use the change-of-base formula.
4 Solve applied problems using common or natural logarithms.

Key Terms
Use the most appropriate term from the given list to complete each statement in exercises 1-3.

$$10 \qquad \log_e x \qquad e \qquad x \qquad \log_{10} x \qquad e^x$$

1. A common logarithm is a logarithm to the base _____.

2. A natural logarithm is a logarithm to the base _____.

3. The inverse of the natural exponential function $f(x) = e^x$ is
    $f^{-1}(x) = $ _____.

Objective 1   Evaluate common logarithms.

Use a calculator to approximate each logarithm to four decimal places.

1. $\log 58$ 1._____

2. $\log 0.482$ 2._____

3. $\log 0.395$ 3._____

4. $\log 40$ 4._____

Evaluate.

5. $\log\left(\dfrac{1}{1000}\right)$

5._____

6. $\log 0.0001$

6._____

7. $\log \sqrt[7]{10,000}$

7._____

8. $\log 10^z$

8._____

9. $\log 0.001$

9._____

Objective 2 Evaluate natural logarithms.

Use a calculator to approximate each logarithm to four decimal places.

10. $\ln 87$

10._____

11. $\ln 45$

11._____

12. $\ln 0.0787$

12._____

Evaluate.

13. $\ln\left(e^{-3}\right)$

13._____

14. $\ln 1$

14._____

15. $\ln \sqrt[10]{e^3}$

15._____

16. $\ln \sqrt[3]{e^5}$

16._____

17. $e^{\ln 18}$

17._____

18. $e^{\ln 23}$

18._____

Objective 3 Use the change-of-base formula.

Use a calculator to approximate each logarithm to four decimal places.

19. $\log_8 80$

19._____

20. $\log_9 14$

20._____

21. $\log_6 0.19$

21._____

22. $\log_2 0.06$

22._____

Objective 4 Solve applied problems using common or natural logarithms.

Solve.

23. The pH of a substance can be determined using the
formula $pH = -\log\left[H^+\right]$, where H^+ represents the
hydrogen ion concentration. To the nearest tenth,
find the pH of the hydrogen ion concentration
4.1×10^{-4}.

23._____

Chapter 9 EXPONENTIAL AND LOGARITHMIC FUNCTIONS

9.6 Exponential and Logarithmic Equations

Learning Objectives
1   Solve exponential equations.
2   Solve logarithmic equations.
3   Solve applied problems using exponential or logarithmic equations.

Objective 1 Solve exponential equations.

Solve. Round each answer to four decimal places.

1. $3^x = 29$ 1._____

2. $5^x = 31$ 2._____

3. $3^x = \dfrac{2}{3}$ 3._____

4. $7^x = \dfrac{2}{7}$ 4._____

5. $3^{2x} = 10$ 5._____

6. $4^{2x} = 11$ 6._____

7. $7^{-x+4} = 53$ 7._____

8. $3^{-x+3} = 74$ 8._____

9. $e^{0.006x} = 35$ 9._____

10. $e^{-0.04x} = 2.2$ 10._____

Objective 2 Solve logarithmic equations.

Solve.

11. $\log_6 (6x - 6) = 1$ 11._____

12. $\log_4 (4x - 7) = 4$ 12._____

13. $\log (x^2 - 1434) = 1$ 13._____

14. $\log (x^2 - 26) = 1$ 14._____

15. $\log_4 x - \log_4 5 = 2$ 15._____

16. $\log_3 x - \log_3 2 = 3$ 16._____

17. $\log_5 x + \log_5 (x - 2) = \log_5 8$ 17._____

18. $\log_3 (x + 3) + \log_3 (x - 3) = 3$ 18._____

19. $\log_3 (x + 1) - \log_3 x = 3$ 19._____

20. $\log_6 x + \log_6 (x - 5) = 2$ 20._____

Objective 3 Solve applied problems using exponential or logarithmic equations.

Solve.

21. How much time will be needed for $22,000 to grow   21._____
to $29,038.45 if deposited at 7% compounded
quarterly? Use the formula $A = P\left(1 + \dfrac{r}{k}\right)^{kt}$. Round
to the nearest tenth.

22. The pH of a substance can be determined using the
formula $pH = -\log[H^+]$, where H^+ represents the
hydrogen ion concentration. The pH of a fruit juice
is 3.3. What is the hydrogen ion concentration of
the juice?

22._____

23. The total expenditures in millions of current dollars
for pollution abatement and control during the
period from 1980 through 1999 can be
approximated by the function $P(x) = 71,897e^{0.055x}$,
where $x = 0$ corresponds to 1980, $x = 1$ to 1981, and
so on. In what year would the total expenditures
reach 128,400 million dollars?

23._____

234

Chapter 10 CONIC SECTIONS

10.1 Introduction to Conics; The Parabola

Learning Objectives
1   Graph a parabola.
2   Write the equation of a parabola in standard form.
3   Solve applied problems involving parabolas.

Key Terms
Use the most appropriate term or phrase from the given list to complete each statement in exercises 1-2.

> **standard form**   **vertex form**   **left or right**   **up or down**

1.   The equation of a parabola with axis of symmetry $x = h$ is in _____
     if it is written as $y = a(x - h)^2 + k$, where (h, k) is the vertex.

2.   The parabola $x = a(y - k)^2 + h$ with vertex (h, k) and axis of symmetry $y = k$ opens to
     the _____.

Objective 1   Graph a parabola.

Identify the vertex and axis of symmetry of each parabola. Then graph.

1. $y = (x + 1)^2 - 3$

1.

2. $y = (x+3)^2 - 2$

2._____

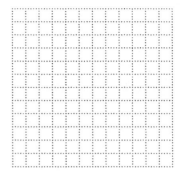

3. $y = -\dfrac{1}{5}(x+4)^2 + 3$

3._____

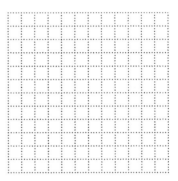

4. $x = 5(y-1)^2$

4._____

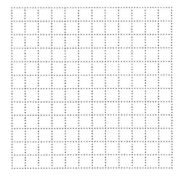

5. $x = -y^2 + 2y + 2$ **5.**_____

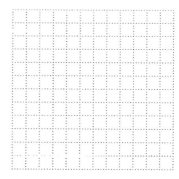

6. $x = -y^2 + 6y + 1$ **6.**_____

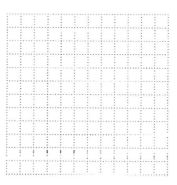

7. $y = 3x^2 + 18x + 23$ **7.**_____

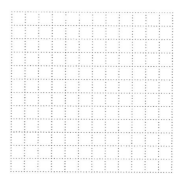

8. $x = -2y^2 - 4y - 3$

8._____

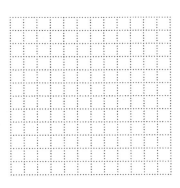

Objective 2 Write the equation of a parabola in standard form.

Write each equation in standard form. Then identify the vertex of each parabola.

9. $y = x^2 - 8x + 19$

9._____

_____

10. $y = x^2 - 10x + 28$

10._____

_____

11. $y = 2x^2 - 9x + 7$

11._____

_____

12. $x = -2y^2 - 16y - 8$

12._____

_____

13. $x = -4y^2 - 24y - 33$

13. _____

_____

14. $x = 5y^2 + 15y + 13$

14. _____

_____

Find the equation in standard form of the parabola whose graph is shown.

15.

15. _____

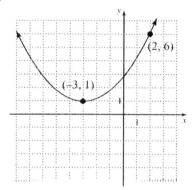

16.

16. _____

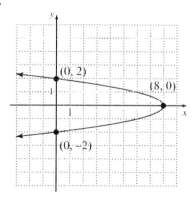

Objective 3 Solve applied problems involving parabolas.

Solve.

17. A farmer has 200 feet of fencing to enclose a
    grazing field. Write an equation in standard form
    that represents the area, A, of the enclosed field in
    terms of the width, w, of the field.

17._____

18. A laboratory designed a radio telescope with a
    diameter of 280 ft and a maximum depth of 42 ft.
    The graph depicts a cross section of this telescope.
    Find the equation of this parabola.

18._____

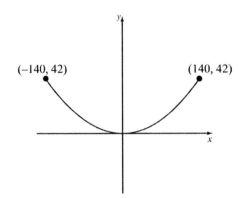

Chapter 10 CONIC SECTIONS

10.2 The Circle

Learning Objectives
1   Use the distance formula and the midpoint formula.
2   Graph a circle.
3   Find the equation of a circle.
4   Write the equation of a circle in standard form.
5   Solve applied problems involving circles.

Key Terms
Use the most appropriate term or phrase from the given list to complete each statement in exercises 1-2.

the radius   any point on the circle   the center   midpoint   distance

1.   In the equation for a circle $(x-h)^2 + (y-k)^2 = r^2$, (h,k) represents

_____.

2.   The _____ between points (x_1, y_1) and (x_2, y_2) is given by

$\sqrt{(x_2 - x_1)^2 + (y_2 - y_1)^2}$.

Objective 1   Use the distance formula and the midpoint formula.

Find the distance between the two points on the coordinate plane. Give an exact answer and also an answer rounded to the nearest tenth.

1. $(3,4)$ and $(10,10)$ 1._____

2. $(6,-14)$ and $(-15,-20)$ 2._____

3. $(13,4)$ and $(6,3)$ 3._____

4. $(1.4,-1.2)$ and $(-6.1,-2.9)$ 4._____

5. $\left(\dfrac{8}{5},\dfrac{1}{10}\right)$ and $\left(\dfrac{6}{5},\dfrac{17}{10}\right)$ 5._____

6. $\left(\dfrac{8}{7},\dfrac{1}{14}\right)$ and $\left(\dfrac{2}{7},\dfrac{19}{14}\right)$ 6._____

Find the coordinates of the midpoint of the line segment joining the points.

7. $(4,-8)$ and $(4,1)$ 7._____

8. $(-2,0)$ and $(1,7)$ 8._____

9. $(-3,-4)$ and $(6,1)$ 9._____

10. $(6.2,-5.8)$ and $(-7.7,-6)$ 10._____

11. $\left(\dfrac{3}{7}, -\dfrac{1}{9}\right)$ and $\left(-\dfrac{1}{2}, \dfrac{3}{2}\right)$ **11.**_____

12. $\left(\dfrac{3}{4}, -\dfrac{4}{5}\right)$ and $\left(-\dfrac{5}{8}, \dfrac{1}{4}\right)$ **12.**_____

Objective 2 Graph a circle.

Find the center and radius of each circle. Then graph.

13. $x^2 + y^2 = 9$ **13.**_____

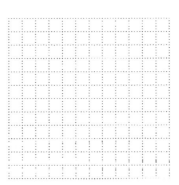

14. $(x+4)^2 + (y+1)^2 = 4$ **14.**_____

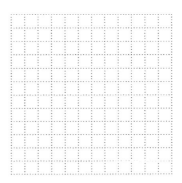

15. $(x-3)^2 + (y+2)^2 = 9$ 15._____

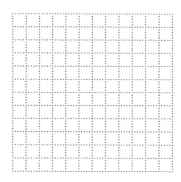

16. $x^2 + y^2 - 8x + 2y - 19 = 0$ 16._____

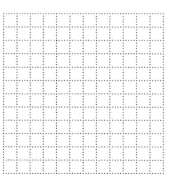

Objective 3 Find the equation of a circle.
Objective 4 Write the equation of a circle in standard form.

Use the given information to find the equation of each circle.

17. Center: $(6,4)$; radius: 5   17._____

18. Center: $(-2,4)$; radius: $\sqrt{3}$ 18._____

19. Center: $(-4,2)$; passes through the point $(-3,5)$ **19.**_____

20. Endpoints of a diameter: $(-2,-5)$ and $(-1,4)$ **20.**_____

Find the center and radius of each circle.

21. $x^2 + y^2 - 12x = 0$ **21.**_____

22. $x^2 + y^2 + 2x - 8y - 28 = 0$ **22.**_____

23. $x^2 + y^2 + 4x - 2y - 235 = 0$ **23.**_____

Objective 5 Solve applied problems involving circles.

Solve.

24. A lawyer drives from her home, located 8 miles east
and 1 mile north of the town courthouse, to her
office, located 6 miles west and 47 miles south of
the courthouse. Find the distance between the
lawyer's home and her office.

24._____

25. A radio station emits a signal that can be received
by a radio within a 17-mile radius of the station. If
the radio station is located 7 miles west and 15
miles north of the center of the city, can a student
whose apartment is located 2 miles east and 2 miles
south of the center of the city listen to the station's
broadcast on his radio?

25._____

Chapter 10 CONIC SECTIONS

10.3 The Ellipse and the Hyperbola

Learning Objectives
1   Graph an ellipse with center at the origin.
2   Graph a hyperbola with center at the origin.
3   Solve applied problems involving ellipses or hyperbolas.

Key Terms
Use the most appropriate term or phrase from the given list to complete each statement in exercises 1-2.

hyperbola   ellipse   circle   asymptotes   axes

1.   A _____ has two branches that extend indefinitely on both sides.

2.   To graph a hyperbola, find the *x*- or *y*-intercepts and the equations of the

_____.

Objective 1   Graph an ellipse with center at the origin.

Graph each ellipse.

1. $\dfrac{x^2}{4} + \dfrac{y^2}{9} = 1$

1.

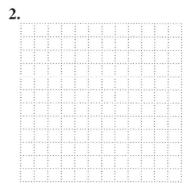

2. $\dfrac{x^2}{9} + \dfrac{y^2}{16} = 1$

2.

3. $\dfrac{x^2}{9} + \dfrac{y^2}{4} = 1$

3.
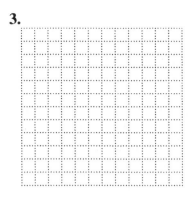

4. $16x^2 + 4y^2 = 64$

4.
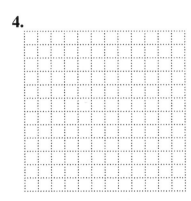

5. $x^2 + 4y^2 = 4$

5.
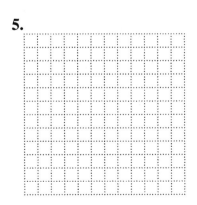

6. $16x^2 + 9y^2 = 144$

6.

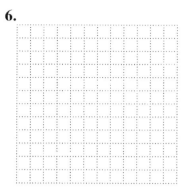

Objective 2   Graph a hyperbola with center at the origin.

Graph each hyperbola.

7. $\dfrac{x^2}{4} - \dfrac{y^2}{9} = 1$

7.

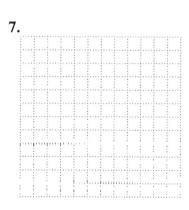

8. $\dfrac{x^2}{4} - \dfrac{y^2}{25} = 1$

8.

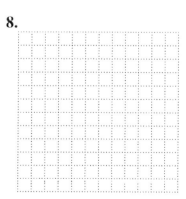

9. $\dfrac{y^2}{9} - \dfrac{x^2}{4} = 1$

9.

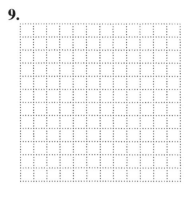

10. $\dfrac{y^2}{1} - \dfrac{x^2}{4} = 1$

10.

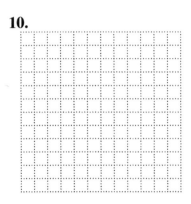

11. $9x^2 - 25y^2 = 225$

11.

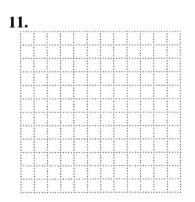

Identify whether the conic section whose equation is given is a parabola, a circle, an ellipse, or a hyperbola.

12. $x^2 + y^2 = 16$ 12._____

13. $4x^2 = 4 - y$ 13._____

14. $x^2 - y^2 = 4$ 14._____

Find the equation of each conic in standard form using the given information.

15. An ellipse centered at the origin that passes through   15._____
the points $(-8,0)$, $(8,0)$, $(0,-3)$, and $(0,3)$.

16. A hyperbola centered at the origin that passes   16._____
through the points $(-3,0)$ and $(3,0)$ and whose

graph approaches the asymptotes $y = -\dfrac{1}{3}x$ and

$y = \dfrac{1}{3}x$.

Objective 3   Solve applied problems involving ellipses or hyperbolas.

Solve.

17. An arch has the shape of half an ellipse. The arch is   17._____
17 meters high at its center and 42 meters wide
across the bottom. Find the equation that best
describes the ellipse.

Chapter 10 CONIC SECTIONS

10.4 Solving Nonlinear Systems of Equations

Learning Objectives
1   Solve a nonlinear system by substitution.
2   Solve a nonlinear system by elimination.
3   Solve applied problems involving nonlinear systems.

Objective 1 Solve a nonlinear system by substitution.

Solve.

1. $y = x^2 - 18$ 1._____
     $y = 3x$

2. $y = x^2 - 35$ 2._____
     $y - 2x$

3. $x^2 + y^2 = 58$ 3._____
     $y - x = 4$

4. $\begin{aligned} x^2 + y^2 &= 117 \\ y - x &= 3 \end{aligned}$

5. $\begin{aligned} x^2 + y^2 &= 9 \\ y^2 &= x + 3 \end{aligned}$

5._____

6. $\begin{aligned} x^2 + y^2 &= 81 \\ y^2 &= x + 9 \end{aligned}$

6._____

7. $\begin{aligned} x^2 + y^2 &= 16 \\ y^2 - 2x^2 &= 10 \end{aligned}$

7._____

8. $\begin{aligned} x^2 + y^2 &= 24 \\ y^2 - 3x^2 &= 12 \end{aligned}$

8._____

9. $\begin{aligned} x^2 + y^2 &= 20 \\ 5x^2 + 3y^2 &= 36 \end{aligned}$

9._____

10. $x^2 + y^2 = 12$
$2x^2 + 3y^2 = 56$

Objective 2 Solve a nonlinear system by elimination.

Solve.

11. $x^2 = 30 - y^2$
$5x^2 = 6 - 2y^2$

11._____

12. $x^2 = 10 - y^2$
$6x^2 = 6 - 3y^2$

12._____

13. $5x^2 - 5y^2 = 485$
$3x^2 - 5y^2 = 285$

13._____

14. $5x^2 - 3y^2 = 230$
$3x^2 - 3y^2 = 132$

14._____

15. $\dfrac{x^2}{4} + \dfrac{y^2}{64} = 1$

$\dfrac{x^2}{2} + \dfrac{y^2}{96} = 1$

15._____

16.
$$\frac{x^2}{4} + \frac{y^2}{16} = 1$$
$$\frac{x^2}{2} + \frac{y^2}{24} = 1$$

Objective 3 Solve applied problems involving nonlinear systems.

Solve.

17. A garden contains two square peanut beds. Find the length of each bed if the sum of the areas is 932 ft^2 and the difference of the areas is 420 ft^2.

17._____

Chapter 10 CONIC SECTIONS

10.5 Solving Nonlinear Inequalities and Nonlinear Systems of Inequalities

Learning Objectives
1   Graph a nonlinear inequality in two variables.
2   Solve a nonlinear system of inequalities by graphing.
3   Solve applied problems involving nonlinear inequalities or nonlinear systems of inequalities.

Objective 1   Graph a nonlinear inequality in two variables.

Graph the solutions of each nonlinear inequality.

1. $x^2 + y^2 > 64$

1.

2. $(x-1)^2 + (y-3)^2 \leq 9$

2.

3. $y > -x^2 + 4x - 8$

3.

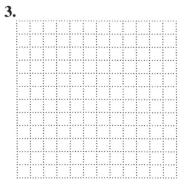

4. $x^2 + 10y^2 > 10$

4.

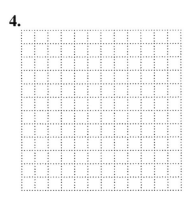

5. $4x^2 + 49y^2 < 196$

5.

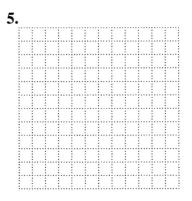

6. $x^2 - y^2 < 9$

6.

Objective 2 Solve a nonlinear system of inequalities by graphing.

Graph the solutions of each nonlinear system of inequalities.

7. $\begin{array}{l} y > x^2 - 4 \\ y \leq -x + 3 \end{array}$

7.

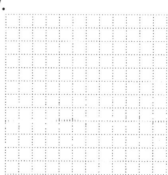

8. $\begin{array}{l} x^2 + y^2 \geq 1 \\ x^2 + y^2 < 4 \end{array}$

8.

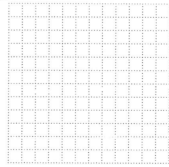

9. $y < x^2 + 4x + 4$

$y > -x^2 - 4x - 7$

9.

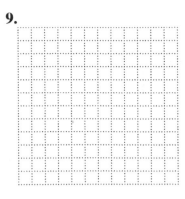

10. $x^2 - y^2 > 9$

$x^2 + y^2 \leq 36$

10.

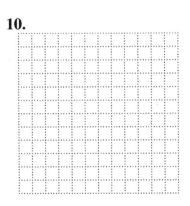

Objective 3   Solve applied problems involving nonlinear inequalities or nonlinear systems of inequalities.

Solve.

11. A search-and-rescue team maps out an elliptical search region to find some hikers. The region is represented by the inequality $16x^2 + 9y^2 \leq 144$. Graph the region.

11.

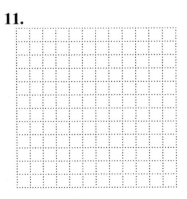

12. Radio station A emits a signal that can be received within a 20-mile radius of the station's transmitting tower. Radio station B located 50 mi away emits a signal that can be received within a 40-mile radius of its tower. The system of inequalities can be used to determine the region in which both radio station signals can be received. Find this region.

$$x^2 + y^2 \le 400$$

$$(x - 50)^2 + y^2 \le 1600$$

12.

Chapter 1 ALGEBRA BASICS

1.1 Introduction to Real Numbers

Key Terms

1. inequality   2. real numbers   3. rational numbers

4. absolute value   5. integers

Objective 1

1. 0, 80   3. $-14, 0, 80, 8.74, 8\frac{1}{2}, -\frac{18}{19}$

5. $-14, 0, 80, 8.74, \sqrt{13}, 8\frac{1}{2}, -\frac{18}{19}$

Objective 2

7.   9.

Objective 3

11. 63   13. 26.3   15. $\frac{12}{7}$

Objective 4

17. false   19. <   21. >

23. $-\frac{5}{2}, -\frac{4}{5}, 3, 4\frac{1}{2}$

25. $[-4,3)$

Objective 5

27. $x < y$ or $y > x$

1.2 Operations with Real Numbers

Key Terms

1. difference   2. negative   3. reciprocal

4. order of operations rule   5. opposite

Objective 1

1. -36 3. -98 5. -10 7. -3.2 9. $-\dfrac{2}{12} = -\dfrac{1}{6}$

11. 20   13. 2   15. $\dfrac{4}{7}$ 17. 16   19. 16   21. $-\dfrac{62}{3}$

23. 7   25. $\sqrt{3}$

Objective 2

27. 781 ft   29. 11,489 ft

1.3 Properties of Real Numbers

Key Terms

1. associative property   2. commutative property   3. distributive property

4. additive inverse property

Objective 1

1. $7 + n$ 3. $5y + 5 \cdot 2$ 5. additive identity property

7. commutative property of addition   9. -4.4 11. 0

13. 5   15. $p \cdot 3r + q \cdot 3r = 3pr + 3qr$

Objective 2

17. Let n represent the number of guests. Then $\dfrac{1}{4}n$ represents the number of guests that

can be served by one recipe. Since $4\left(\dfrac{1}{4}\right)n = \left(4 \cdot \dfrac{1}{4}\right)n = n$ by the multiplication inverse property, Elise must multiply the ingredients in the recipe by 4 to make sure she has enough dessert for her n guests.

1.4 Laws of Exponents and Scientific Notation

Key Terms

1. standard notation   2. divided   3. zero   4. added

Objective 1

1. 1   3. -2 5. $\dfrac{1}{10^2}$ 7. $\dfrac{x^5}{y^2}$

Objective 2

9. 7^{15} 11. cannot be simplified   13. y^{13}

Objective 3

15. m^{28} 17. $\dfrac{a^8}{b^8}$ 19. $-\dfrac{f^{15}}{g^{30}}$

Objective 4

21. 4.8×10^{-10} 23. $46,000,000$ 25. 0.0000134 27. 2.508×10^{-1}

Objective 5

29. $\$1.3 \times 10^7$

1.5 Algebraic Expressions: Translating, Evaluating, and Simplifying

Key Terms

1. exponent   2. terms   3. variables   4. expression

Objective 1

1. three plus twice s 3. the product of seven and the difference between x and y

5. $\dfrac{1}{4}n$ 7. $\dfrac{c}{d}$ 9. $2x + 7$

Objective 2

11a. 7   b. 9   c. 5   13. 0   14. -21 15. -5

Objective 3

17. like   19. $11t$ 21. cannot be combined   23. $-4t + 2$

25. $19m - 54$

Objective 4

27. $7x + 12$ 29. $44 + y$

Chapter 2 LINEAR EQUATIONS AND INEQUALITIES

2.1 Solving Linear Equations

Key Terms

1. equivalent equations   2. distributive property   3. addition property of

equality   4. equation

Objective 1

1. not a solution   3. solution   5. not a solution

Objective 2

7. 17   9. -7

Objective 3

11. 44   13. -9 15. $-\dfrac{2}{15}$

Objective 4

17. -160 19. -3

Objective 5

21. 5   23. $-\dfrac{4}{3}$ 25. $\dfrac{1}{2}$

Objective 6

27. 43 degrees, 43 degrees, 94 degrees   29. 200 kg

2.2 Solving Literal Equations and Formulas

Objective 1

1. $x = z + 7b$ 3. $n = \dfrac{s - 3v}{3}$ 5. $v_1 = v_2 - ta$ 7. $M = \dfrac{B}{W}$

9. $k = \dfrac{5}{7}C + 48$ 11. $w = \dfrac{V}{lh}$ 13. $c = \dfrac{S - 2ab}{2a + 2b}$

Objective 2

15. 6 inches

2.3 Solving Linear Inequalities

Key Terms

1. positive   2. less than   3. equation   4. excludes

5. distributive property   6. multiplication property

Objective 1

1. not a solution   3. not a solution

Objective 2

5. $(-\infty, 3)$

7. $(7, \infty)$

Objective 3

9. $x \ge 6$; interval notation: $[6, \infty)$

11. $x \ge 3$; interval notation: $[3, \infty)$

13. $x \leq 4$; interval notation: $(-\infty, 4]$

15. $x \geq 4$; interval notation: $[4, \infty)$

17. $x > 3$; interval notation: $(3, \infty)$ 19. $x > -\dfrac{2}{25}$; interval notation: $\left(-\dfrac{2}{25}, \infty\right)$

Objective 4

21. You can drive at most 250 mi per day.   23. He can invest at most $3000.

2.4 Solving Compound Inequalities

Key Terms

1. intersection   2. union   3. inequalities   4. or

Objective 1

1. $-2 < x < 6$; interval notation: $(-2, 6)$

3. $-2 < y < 2$; interval notation: $(-2, 2)$

5. $-6 < x < 2$; interval notation: $(-6, 2)$

7. $-6 < x \leq 6$; interval notation: $(-6, 6]$ 8. no solution   9. no solution

Objective 2

11. $x \geq -4$; interval notation: $[-4, \infty)$

13. $x < -2$ or $x > 4$; interval notation: $(-\infty, -2) \cup (4, \infty)$

15. $x < 5$ or $x > 9$; interval notation: $(-\infty, 5) \cup (9, \infty)$

17. $x > 0$; interval notation: $(0, \infty)$ 19. all real numbers;

interval notation: $(-\infty, \infty)$

Objective 3

21. Jerome's body mass index will fall between 23 and 32 for the set of all weights such that $148 < W < 205$.   23. Telephoning time must be between 789 minutes and 875 minutes.

2.5 Solving Absolute Value Equations and Inequalities

Key Terms

1. plus or minus   2. no solutions   3. one solution

4. the opposite of that number.

Objective 1

1. $-19, 19$ 3. $-8, 8$ 5. $-6, 4$ 7. no solution

9. $-25, \dfrac{87}{13}$

Objective 2

11. $-1 < x < 1$; interval notation: $(-1,1)$

13. $-15 \le x \le -5$; interval notation: $[-15, -5]$

15. $x < -33$ or $x > -1$;

interval notation: $(-\infty, -33) \cup (-1, \infty)$

17. $\dfrac{1}{4} \le x \le \dfrac{7}{4}$; interval notation: $\left[\dfrac{1}{4}, \dfrac{7}{4}\right]$

19. $\dfrac{3}{4} \le x \le \dfrac{21}{4}$; interval notation: $\left[\dfrac{3}{4}, \dfrac{21}{4}\right]$

Objective 3

21. Kathy weighs between 151 and 163 pounds.

Chapter 3 GRAPHS, LINEAR EQUATIONS AND INEQUALITIES, AND FUNCTIONS

3.1 The Rectangular Coordinate System

Key Terms

1. horizontal   2. *y*-axis   3. Quadrant II   4. second

5. *x*-axis

Objective 1

1.   3.

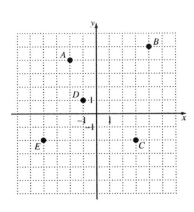   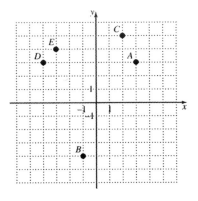

5. $A(2,2)$, $B(5,2)$, $C(-4,2)$, $D(3,-4)$, $E(4,-3)$

Objective 2

7. II   9. III   11. II   13. III   15. I

Objective 3

17. The sales for 1995 were $17.5 million.

3.2 Slope

Key Terms

1. increasing   2. undefined   3. negative   4. 0

Objective 1

1. $m = \dfrac{7}{3}$

3. undefined slope

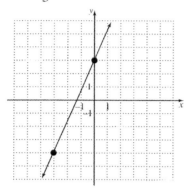

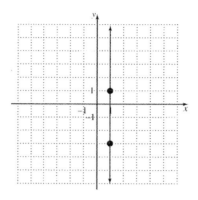

5. slope of $\overrightarrow{AB}$: -1; slope of $\overrightarrow{CD}$: $\dfrac{3}{2}$

7.

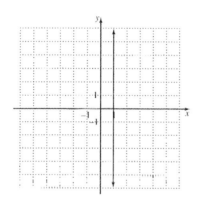

Objective 2

9. zero slope   11. positive slope

Objective 3

13. parallel   15. perpendicular

Objective 4

17. The total number of miles is increasing.

3.3 Graphing Linear Equations

Key Terms

1. *x*-axis   2. *x*-intercept   3. solution   4. table

Objective 1

1. not a solution   3. solution   5. not a solution

Objective 2

7.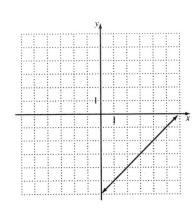

9.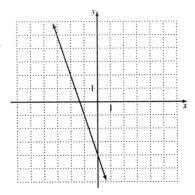

11.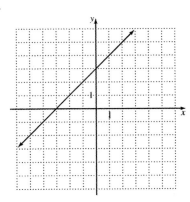

13.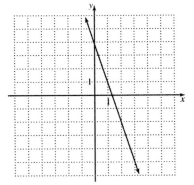

15.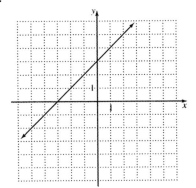

Objective 3

17. x-intercept: $(12,0)$; y-intercept: $(0,-15)$ 19. x-intercept: none;

y-intercept: $(0,-2)$

Objective 4

21a. $P = 30m + 20$ b.   50; 80; 110   c.

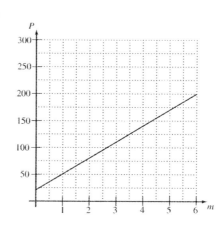

3.4 More on Graphing Linear Equations

Key Terms

1.   point-slope form   2. y-intercept   3.   one point on the line

4.   slope-intercept form

Objective 1

1.

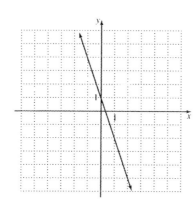

3.

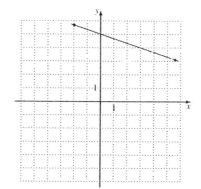

Answers to Worksheets for Classroom or Lab Practice

5. 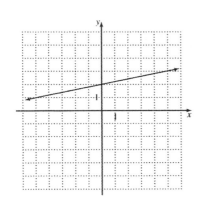   7.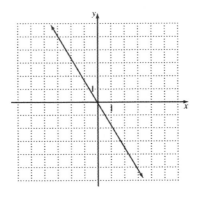

Objective 2

9. $y = 3x$; slope: 3; y-intercept: $(0,0)$ 11. $y = -\dfrac{6}{5}x - 2$; slope: $-\dfrac{6}{5}$;

y-intercept: $(0,-2)$ 13. $y = \dfrac{1}{2}x + \dfrac{5}{2}$; slope: $\dfrac{1}{2}$; y-intercept: $\left(0,\dfrac{5}{2}\right)$

15. $y = 2x - 15$; slope: 2; y-intercept: $(0,-15)$

Objective 3

17. $y + 4 = \dfrac{2}{3}(x-3)$ or $y + 6 = \dfrac{2}{3}(x-0)$

Objective 4

19. $y - 9 = \dfrac{1}{2}(x-4)$ 21. $y - \dfrac{2}{3} = -\dfrac{5}{3}\left(x - \dfrac{1}{3}\right)$

Objective 5

23a. $C = 4x + 300$ b.   The cost of producing 25 shirts is $400.

c.

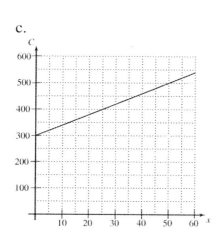

3.5 Graphing Linear Inequalities

Key Terms

1. test point   2. broken   3. linear inequality   4. is not

Objective 1

1. solution   3. solution   5. not a solution

7.   9.

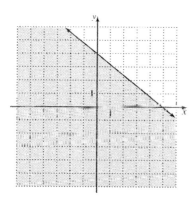   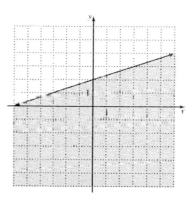

11.   13.

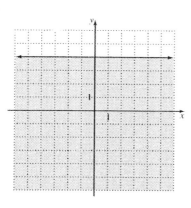   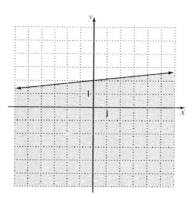

15.

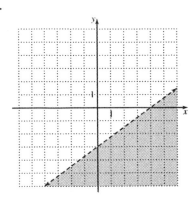

17.

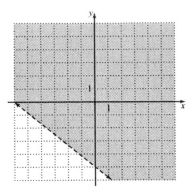

19.

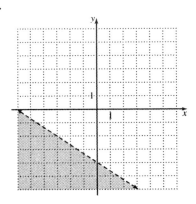

Objective 2

21a. $5x + 8y > 50$ b.

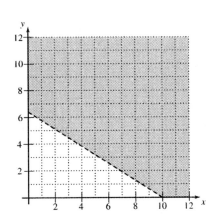

3.6 Introduction to Functions

Key Terms

1. relation   2. represents   3. function   4. domain

5. range   6. dependent

Objective 1

  1.   function   3.   not a function

Objective 2

  5.   domain: $\{-4,-2,0,2,3\}$; range: $\{-64,-8,0,8,27\}$

  7.   domain: $\{-4,-3,0,2,4\}$; range: $\{-1,0,2,4,5\}$

Objective 3

  9a. -8 b. 4 c. -4 d. -1.4 11a. 9 b. 0

  c. $|t-9|$ d. $|t-10|$ 13a. -7 b. -7 c. -7 d. -7

Objective 4

  15.   domain: all real numbers; range: all real numbers

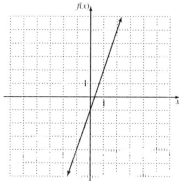

  17.   domain: all real numbers; range: all real numbers

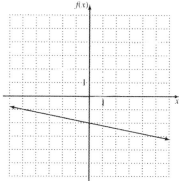

Objective 5

  19.   not a function   21.   function

Objective 6

23a. $C(t) = 20t + 40$ b.   c.   The total cost is $240.

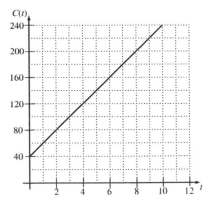

d. $t \geq 0$

Chapter 4 SYSTEMS OF LINEAR EQUATIONS AND INEQUALITIES

4.1 Solving Systems of Linear Equations by Graphing

Key Terms

1. dependent   2. true   3. graph   4. independent

Objective 1

1. solution   3. solution

Objective 2, Objective 3

5. The solution is $(2, 4)$.   7. The solution is $(-2, 1)$.

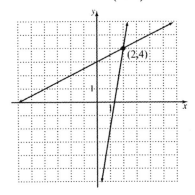 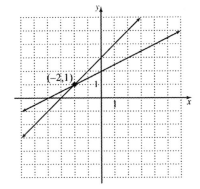

280

9. The solution is $(1,0)$.

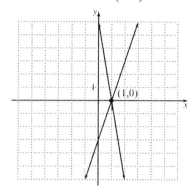

11. Infinitely many solutions

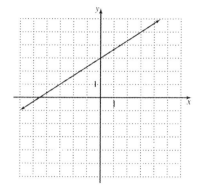

13. The solution is $(-1.8, 7.4)$.

Objective 4

15a. Let w represent the wife's income; let h represent the husband's income.

$$w + h = 67,000$$

$$w = h + 3000$$

b.

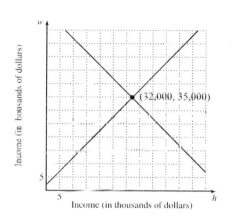

c. The wife's income is $35,000 and the husband's income is $32,000.

17a. $\begin{array}{l} p + l = 1600 \\ 25p + 20l = 36{,}000 \end{array}$

b.

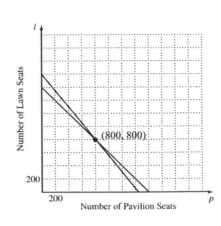

c.   800 pavilion seats and 800 lawn seats were sold.

4.2 Solving Systems of Linear Equations Algebraically by Substitution or Elimination

Key Terms

1.   true   2.   false   3.   original   4.   substitution

Objective 1

1. $(6, 2)$ 3.   infinitely many solutions   5.   No solution

Objective 2

7. $(9, 2)$ 9.   no solution   11. $(8, 6)$

Objective 3

13.   The trains meet 300 km from the station.   15.   Mix 14 oz of lemon juice with

28 oz of food-grade linseed oil.   17.   11 L of the 20% juice should be mixed with

4 L of the 5% juice.

4.3 Solving Systems of Linear Equations in Three Variables

Objective 1

1.   not a solution   2.   not a solution   3.   solution

Objective 2

5. $(0,-5,5)$ 7. $(-5,-4,3)$ 9. $(-5,-1,6)$ 11. $(-3,6,-5)$

Objective 3

13. The lengths are 5, 19, and 23 cm.   15. The first part of the investment was

$54,000, the second part was $12,000, and the third part was $20,000.

4.4 Solving Systems of Linear Equations by Using Matrices

Key Terms

1. constants   2. columns   3. augmented   4. rows

Objective 1

1. $(6,-4)$ 3. $(3,-2)$ 5. $(-2,13,10)$ 7. no solution

9. infinitely many solutions

Objective 2

11. The grocer should use 12.6 lb of candy.

4.5 Solving Systems of Linear Inequalities

Key Terms

1. shaded regions   2. boundary line.

Objective 1

1.

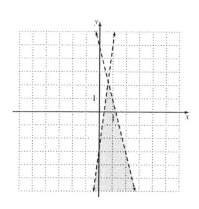

3.

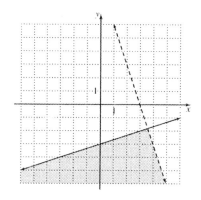

5.

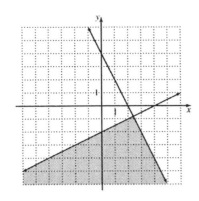

7.

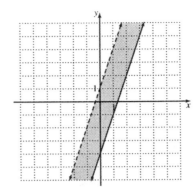

9.

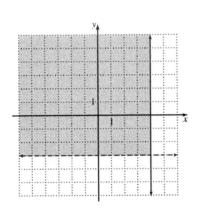

11.

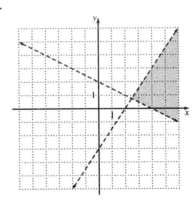

13.

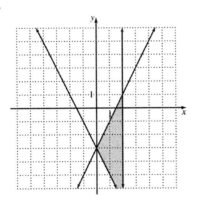

15.

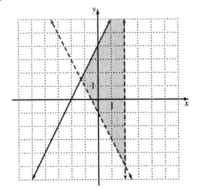

Objective 2

17a. $\begin{aligned} x + y &\le 7 \\ x + 3y &\le 15 \\ x &\ge 0 \\ y &\ge 0 \end{aligned}$

b.

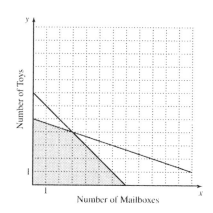

c. The integer solutions in this region represents all possible numbers of mailboxes and toys that Carlo and Anita can produce given their time restrictions.

Chapter 5 POLYNOMIALS

5.1 Addition and Subtraction of Polynomials

Key Terms

1. polynomial   2. with two terms   3. leading coefficient

4. polynomial

Objective 1

1. polynomial   3. not a polynomial   5. terms; $6x^3, -3x^2, 4x, -5$;

coefficients: $6, -3, 4, -5$ 7. terms: $2x^2y^3, -3x^3y^2$; coefficients: $2, -3$

9. polynomial of degree 4   11. binomial of degree 9

13. $-7y^4 + 6y^2 + 7y - 3$; leading term: $-7y^4$; leading coefficient: -7

Objective 2

15. $6b^5 - b^3 - 4b^2$

Objective 3

17. 37

Answers to Worksheets for Classroom or Lab Practice

Objective 4

19. $x^2 - 2x - 3$ 21. $7x^4 + 9x^3 - 7x^2 + 8x - 14$ 23. $15f^3 - 3f^2 + 3f - 9$

25. $-9a + 8$ 27. $f(x) + g(x) = -5x^2 + 10x - 17$;

$f(x) - g(x) = 13x^2 + 4x + 7$

Objective 5

29. The total revenue is $35,454.64.

5.2 Multiplication of Polynomials

Key Terms

1. distributive property   2. outer   3. square of a binomial

4. binomials

Objective 1

1. $16x^{18}$ 3. $64s^6t^6$ 5. $-8x^2 + 24x$ 7. $32x^4 - 12x^3 + 28x^2$

9. $38x^2 + 28x - 452$

Objective 2

11. $x^2 + 9x + 20$ 13. $b^2 + 4b + 4$ 15. $72z^2 + 50z + 8$

17. $16z^2 + 30zu - 4u^2$

Objective 3

19. $x^2 + 22x + 121$ 21. $4s^2 - 9l^2$ 23. $27g^3 - 27g^2 j + 9gj^2 - j^3$

25. $f(x) \cdot g(x) = 8x^4 - 14x^3 - 8x^2 + 21x - 6$ 27. $-63 - 34m - 4m^2$

29. $5h + 2hy + h^2$

5.3 Division of Polynomials

Key Terms

1. quotient   2. remainder

Objective 1

1. $16r^4$ 3. $-5s^3t^2$ 5. $5x^7 - 9x^2$ 7. $3x^3 - 5x^2 + 4$

9. $9v^6 p^9 - 4v^4 p^7 + 4p^3$

Objective 2

11. $2x^2 - 6x - 5$ 13. $7w^2 - 2w + 8 + \dfrac{7}{4w+4}$ 15. $8k^2 + 7k - 2$

17. $x^2 - 10x + 100$ 19. $3k^2 + 7k - 1$

Objective 3

21a. $\dfrac{8.32}{x} + 0.04$ b. The average cost for 1 hr of long-distance calls is $0.18 per

minute.

5.4 The Greatest Common Factor and Factoring by Grouping

Key Terms

1. product   2. distributive property   3. factor

Objective 1

1. $5y\left(4y^3 + 1\right)$ 3. $7x^4\left(x^2 - 4x + 2\right)$ 5. $-4m^2n^2\left(m^5n^3 + 7m^3n^2 + 7\right)$

7. $\left(r+7\right)\left(b-5\right)$ 9. $\left(r+7s^2\right)\left(r^2 - 5s + 1\right)$

Objective 2

11. $\left(p^2 - 3\right)\left(y - t\right)$ 13. $\left(s + f\right)\left(q - 8\right)$ 15. $\left(b + 6\right)\left(b + 3\right)$

17. $\left(2s + 5\right)\left(3t - 5\right)$ 19. $\left(r + 5w\right)\left(r - 2t\right)$ 21. $\left(a^2 + b^2\right)\left(5b + 2a\right)$

Objective 3

23a. $h(t) = 16t(6-t)$ b.   128 ft

5.5 Factoring Trinomials

Key Terms

1. prime   2. the *ac* method

Objective 1

1. $x-9$ 3. $(r+3)(r+7)$ 5. $(w-2)(w-4)$

7. $(s-3)(s+7)$ 9. $(v+4)(v-5)$ 11. $(a+8y)(a-10y)$

13. prime polynomial

Objective 2

15. $4a-3b$ 17. $(7s+3)(8s+1)$ 19. $(5c-6)(c+5)$

21. $(4w-3)(5w+1)$ 23. $(3c-2w)(6c-7w)$ 25. $(7s+9b)(8s-7b)$

27. $(b^3+6)(b^3-10)$ 29. $-2(3m+4)(2m-7)$

5.6 Special Factoring

Key Terms

1. the difference of squares   2.   the difference of two perfect cubes

Objective 1

1. $(s-4)^2$ 3. $(v+3)^2$ 5. $(2c+9)^2$ 7. $(9s-f)^2$

9. $(4s^4-5f)^2$

Objective 2

11. $(v+9)(v-9)$ 13. $(6+v)(6-v)$ 15. $(8x+11)(8x-11)$

17. $\left(6r^2+5\right)\left(6r^2-5\right)$ 19. $(x-v)(p+2)(p-2)$

Objective 3

21. $(c+2)\left(c^2-2c+4\right)$ 23. $(5x+2y)\left(25x^2-10xy+4y^2\right)$

25. $5a^3(a-6)\left(a^2+6a+36\right)$

Objective 4

27. $17,000(r+1)^2$

5.7 Solving Quadratic Equations by Factoring

Key Terms

1. the squares of the legs; the square of the hypotenuse   2. zero

Objective 1

1. $-28,\ 32$ 3. $-\dfrac{3}{5},\ 3$ 5. $-5,\ 0$ 7. $0,\ \dfrac{9}{5}$

9. $-2,\ -7$ 11. $2,\ 8$ 13. $4,\ -2$ 15. $\dfrac{9}{8},\ \dfrac{3}{4}$ 17. $2,\ -2$

19. $0,\ \dfrac{9}{4}$ 21. $2,\ -2$ 23. $4,\ 9$ 25. $-1,\ 4$ 27. $-6,\ -10$

Objective 2

29a. The projectile will reach the ground in 8 sec.   b. The projectile will be 528 ft above the ground after 5 sec.

Chapter 6   RATIONAL EXPRESSIONS AND EQUATIONS

6.1 Multiplication and Division of Rational Expressions

Key Terms

1. rational number   2. undefined   3. common factors

   4.   opposites

Objective 1

   1.   0   3. $-\dfrac{6}{7}$ 5.   5, −1

Objective 2

   7.   equivalent   9. $\dfrac{8w^3x^3}{5}$ 11. $\dfrac{3x+6}{x}$ 13. $\dfrac{3}{3a-7}$ 15. $\dfrac{2}{t+4}$

   17. $\dfrac{z-8}{z+4}$

Objective 3

   19. $\dfrac{3y^2}{x^3}$ 21. $\dfrac{(5z+4)(z+5)}{6z}$ 23. $\dfrac{t-10}{t+10}$

Objective 4

   25. $\dfrac{7a^3b^6}{3}$ 27. $\dfrac{(x+5)(x+3)}{4(x-5)}$ 29. $(k-4)(k-6)$

   31. $f(x)\cdot g(x)=\dfrac{9(x-10)}{(x+8)^2}$; $f(x)\div g(x)=\dfrac{x-10}{9x^2}$

6.2 Addition and Subtraction of Rational Expressions

Key Terms

   1.   denominators   2.   numerators; denominators   3.   simplest form

Objective 1

   1. $\dfrac{6x+14}{x-5}$ 3. $\dfrac{c+2d}{c^2d}$ 5. $\dfrac{1}{n-a}$

Objective 2

   7.   LCD $= 65a^4$; $\dfrac{55a}{65a^4}$ and $\dfrac{52}{65a^4}$ 9.   LCD $= 15(x+1)$; $\dfrac{50}{15(x+1)}$ and

$$\frac{-3}{15(x+1)}$$

11. LCD $= (m+7)^2(m+8)$; $\dfrac{m^2+16m+64}{(m+7)^2(m+8)}$ and

$$\frac{m^2-49}{(m+7)^2(m+8)}$$

Objective 3

13. $\dfrac{49-3z}{63z^3y}$ 15. $\dfrac{n^2+11n+12}{(n+3)(n+7)}$ 17. $\dfrac{6t-13}{(t-3)(t+3)}$

19. $\dfrac{2y^2+y-3}{(y+4)(y+1)(y+3)}$

Objective 4

21. $\dfrac{175+1.95x}{x}$

6.3 Complex Rational Expressions

Key Terms

1. reciprocal   2. rational

Objective 1

1. $72n$ 3. $\dfrac{2}{a}$ 5. $\dfrac{4}{9x-1}$ 7. $\dfrac{3s+1}{s}$ 9. $\dfrac{p-5}{p}$

11. $\dfrac{33(4-a^3)}{14a^2(2a+15)}$ 13. $y-3$ 15. $\dfrac{11z-51}{-z+89}$ 17. $\dfrac{z-7}{z+1}$

19. $\dfrac{m^5+n^3}{mn^3+m^5n}$

6.4 Solving Rational Equations

Key Terms

   1. equal   2. are not   3. LCD

Objective 1

   1. $\dfrac{5}{2}$ 3. 15   5. no solution   7. $4, -4$ 9. 34

   11. -3 13. $P = \dfrac{B}{T}$ 15. 11

Objective 2

   17. 1047 deer are in the preserve.

6.5 Variation

Key Terms

   1. directly proportional   2. inversely

Objective 1

   1. direct variation   3. $k = 8;\ y = 8x$ 5. $k = \dfrac{6}{31};\ y = \dfrac{6}{31}x$

   7. $k = \dfrac{3}{2};\ y = \dfrac{3}{2}x$ 9. $k = 126;\ y = \dfrac{126}{x}$ 11. $k = 0.06;\ y = \dfrac{0.06}{x}$

   13. $k = 1;\ y = \dfrac{1}{x}$ 15. $k = 1;\ y = xz$ 17. $k = 3;\ y = \dfrac{3wx^2}{z}$

Objective 2

   19. The number of cans used is 413,524,000.   21. The pressure is 108.89 pounds

per square foot.

Chapter 7 RADICAL EXPRESSIONS AND EQUATIONS

7.1 Radical Expressions and Rational Exponents

Key Terms

1. square   2. is   3. rational   4. multiplied by

Objective 1

1. 22   3. not a real number   5. 10   7. $-\dfrac{3}{2}$ 9. 1.414

11. 2.668

Objective 2

13. $5x^3$ 15. $8x^4y^9$ 17. $-5x^3$ 19. $5y^2$ 21. -3

23. 36   25. 27   27. $\sqrt[4]{q}$ 29. $4x^{12}y^6$

7.2 Simplifying Radical Expressions

Key Terms

1. multiply   2. is not

Objective 1

1. 2   3. $\sqrt[20]{x}$ 5. $\sqrt[15]{y}$ 7. $\sqrt{30}$ 9. $\sqrt[3]{110wz}$ 11. $3\sqrt{7}$

13. $-2\sqrt[4]{6}$ 15. $8x^3y^7\sqrt{7y}$ 17. 2   19. $2p\sqrt{p}$ 21. $\dfrac{7}{9}$

23. $\dfrac{\sqrt{7}}{x^3}$ 25. $2\sqrt{26}$

Objective 2

27. The height is 18in.

Answers to Worksheets for Classroom or Lab Practice

7.3 Addition and Subtraction of Radical Expressions

Key Terms

1. unlike; different   2. radicands

Objective 1

1. $12\sqrt{3}$ 3. Cannot be combined   5. $12\sqrt{x}$ 7. $9\sqrt[4]{y}$

9. $4\sqrt{x-2}$ 11. $8\sqrt{5}$ 13. $11\sqrt{5}$ 15. $29\sqrt{2}$ 17. $23\sqrt[3]{3}$

19. $(35a+1)\sqrt{2a}$ 21. $(4+3xy)\sqrt[4]{xy^2}$

Objective 2

23. The length of a side of the square with an area of 448 square feet is $8\sqrt{7}$ feet. The

length of a side of the square with an area of 112 square feet is $4\sqrt{7}$ feet. The difference

in length of the two sides is $4\sqrt{7}$ feet.

7.4 Multiplication and Division of Radical Expressions

Key Terms

1. conjugate   2. denominator

Objective 1

1. $3\sqrt{10}$ 3. $12-3\sqrt{6}$ 5. $12\sqrt{255}+540$ 7. 4

9. $10x-25$

Objective 2

11. $4n^7$ 13. $\dfrac{y^4}{7}$ 15. $\dfrac{2x^3}{y^3}\sqrt{5xy}$

Objective 3

17. $\dfrac{\sqrt{6x}}{9}$ 19. $\dfrac{\sqrt{6x}}{x}$ 21. $\dfrac{5-\sqrt{7}}{18}$ 23. $\dfrac{a-2\sqrt{ab}+b}{a-b}$

Objective 4

25. 8 in.^{2}

7.5 Solving Radical Equations

Objective 1

1. 25   3. $\dfrac{43}{5}$ 5. no solution   7. $\dfrac{49}{5}$ 9. 3

11. no solution   13. -10 15. 6   17. no solution   19. 4

Objective 2

21. The car will skid 180 feet.

7.6 Complex Numbers

Key Terms

1. imaginary part   2. complex number

Objective 1

1. $8i$ 3. $4i\sqrt{5}$

Objective 2

5. $15 + i$ 7. $10+6i$ 9. $14-6i$

Objective 3

11. -20 13. $-9-6i$ 15. $90-30i$ 17. $-16+30i$

Objective 4

19. $7-2i$; 53

Objective 5

21. $-\dfrac{7}{9}i$ 23. $\dfrac{23}{53}+\dfrac{1}{53}i$

Objective 6

25. i

Objective 7

27. $(22+2i)$ ohms

Chapter 8 QUADRATIC EQUATIONS, FUNCTIONS, AND INEQUALITIES

8.1 Solving Quadratic Equations by Completing the Square

Objective 1

1. $-7, 7$ 3. $-6i, 6i$ 5. $-\dfrac{\sqrt{34}}{2}, \dfrac{\sqrt{34}}{2}$ 7. $\dfrac{-1-3\sqrt{5}}{5}, \dfrac{-1+3\sqrt{5}}{5}$

9. $\dfrac{1-2i\sqrt{2}}{3}, \dfrac{1+2i\sqrt{2}}{3}$ 11. $r=\dfrac{\pm\sqrt{3\pi xy}}{\pi x}$

Objective 2

13. 1 15. $\dfrac{25}{4}$ 17. $1, 3$ 19. $-1-\sqrt{7}, -1+\sqrt{7}$

21. $\dfrac{-10-\sqrt{109}}{3}, \dfrac{-10+\sqrt{109}}{3}$ 23. $\dfrac{-1-\sqrt{17}}{4}, \dfrac{-1+\sqrt{17}}{4}$ 25. $-\dfrac{3}{4}, 1$

27. $3-\sqrt{13}, 3+\sqrt{13}$

Objective 3

29. The object would fall for 9.2 seconds.

8.2 Solving Quadratic Equations by Using the Quadratic Formula

Key Terms

1. negative   2. quadratic equation

Objective 1

1. $-3, 5$

3. $-5 - \sqrt{21}, -5 + \sqrt{21}$

5. $-9 - \sqrt{77}, -9 + \sqrt{77}$

7. $1 - 6i, 1 + 6i$

9. $\dfrac{1 - \sqrt{21}}{6}, \dfrac{1 + \sqrt{21}}{6}$

11. $\dfrac{1 - i\sqrt{35}}{2}, \dfrac{1 + i\sqrt{35}}{2}$

13. $\dfrac{5}{3}$

15. $1 - \sqrt{15}, 1 + \sqrt{15}$

17. $\dfrac{-1 - \sqrt{589}}{14}, \dfrac{-1 + \sqrt{589}}{14}$

19. $\dfrac{1 - \sqrt{17}}{2}, \dfrac{1 + \sqrt{17}}{2}$

Objective 2

21. two complex solutions (containing i)

23. two real solutions

Objective 3

25. Supply and demand are equal at 70 cents per muffin.

8.3 More on Quadratic Equations

Objective 1

1. $-\dfrac{9}{10}, 1$

3. $-\sqrt{6}, -1, 1, \sqrt{6}$

5. 16

7. 81

9. $-1, 125$

11. $2, 3$

Objective 2

13. $x^2 + 8x - 20 = 0$

15. $5m^2 - 29m + 20 = 0$

17. $49s^2 - 21s - 54 = 0$

19. $x^2 - 20x + 100 = 0$

21. $t^2 - 14 = 0$

23. $x^2 + 4 = 0$

Objective 3

25. The speed on the first part of the trip was 40 mph. The speed on the second part of the trip was 35 mph.

Answers to Worksheets for Classroom or Lab Practice

8.4 Graphing Quadratic Functions

Key Terms

1. minimum   2. downward   3. y-coordinate

Objective 1

1.

x	$y = f(x) = -3x^2$	$(x,\ y)$
-2	-12	$(-2, -12)$
-1	-3	$(-1, -3)$
0	0	$(0, 0)$
1	-3	$(1, -3)$
2	-12	$(2, -12)$

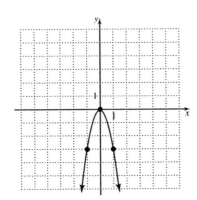

3.

x	$y = f(x) = \dfrac{1}{4}x^2$	(x, y)
-4	4	$(-4,4)$
-2	1	$(-2,1)$
0	0	$(0,0)$
2	1	$(2,1)$
4	4	$(4,4)$

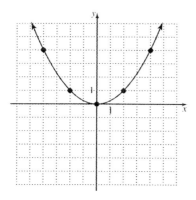

Objective 2

5.   vertex: $(-1,4)$; axis of symmetry: $x = -1$; x-intercepts: $(-3,0)$ and $(1,0)$;

y-intercept: $(0,3)$

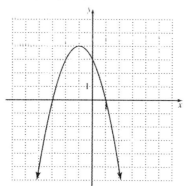

7. vertex: $(0,-16)$; axis of symmetry: $x=0$; x-intercepts: $(-4,0)$ and $(4,0)$;

    y-intercept: $(0,-16)$

    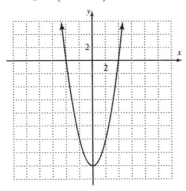

9. vertex: $(-8,0)$; axis of symmetry: $x=-8$; x-intercept: $(-8,0)$; y-intercept: $(0,64)$

    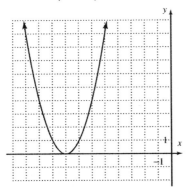

11. vertex: $(2,9)$; axis of symmetry: $x=2$; x-intercepts: $(-1,0)$ and $(5,0)$;

    y-intercept: $(0,5)$

    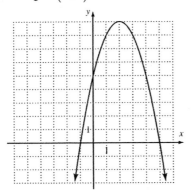

13. domain: $(-\infty, \infty)$; range: $[-22, \infty)$

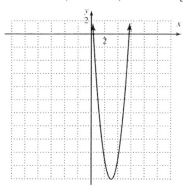

15. domain: $(-\infty, \infty)$; range: $[-6, \infty)$

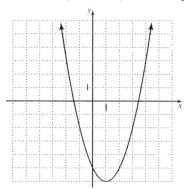

Objective 3

17. The dimensions are 65 ft by 65 ft. The area is 4225 sq ft.

8.5 Solving Quadratic and Rational Inequalities

Key Terms

1. equation   2. boundary points

Objective 1

1. $(-3, 3)$

3. $(-\infty, \infty)$

5. $(-\infty, -3] \cup [8, \infty)$

6. $(3, 7)$

9. $\left(-\infty,-\dfrac{1}{2}\right]\cup\left[\dfrac{1}{7},\infty\right)$

11. $\left(-\infty,-\dfrac{1}{3}\right]\cup\left[\dfrac{1}{5},\infty\right)$

13. $\left(-\infty,7\right)$

15. $\left(-\infty,-6\right)\cup\left[-2,\infty\right)$

17. $\left[-\dfrac{3}{2},6\right)$

Objective 2

19. The rocket is at least 144 ft above the ground from 1.5 sec to 6 sec.

Chapter 9 EXPONENTIAL AND LOGARITHMIC FUNCTIONS

9.1 The Algebra of Functions and Inverse Functions

Key Terms

1. g; $g(x)$; f 2. second   3. horizontal   4. inverse

Objective 1

1. $(f+g)(x)=2x$, $(f-g)(x)=10$, $(f\cdot g)(x)=x^2-25$, $\left(\dfrac{f}{g}\right)(x)=\dfrac{x+5}{x-5}$

3. $\left(\dfrac{f}{g}\right)(x)=\dfrac{4(8-x)}{x+2}$ 5. 38   7. -17

Objective 2

9. $(f\circ g)(x)=18x^2-48x+37$, $(g\circ f)(x)=6x^2+11$

Objective 3

11. one-to-one   13. not one-to-one

Objective 4

15. $\{(5,-9),(-6,1),(9,8),(-5,5)\}$ 17. $g^{-1}(x)=x+3$

19. $f^{-1}(x) = \sqrt[3]{x+1}$ 21. $g^{-1}(x) = \dfrac{7x+3}{1-x}$ 23. inverses

Objective 5

25.

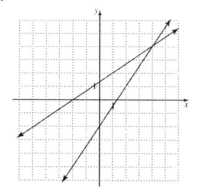

Objective 6

27. $C^{-1}(18)$ represents the number of people that can be taken for a cost of $18 per

person.

9.2 Exponential Functions

Key Terms

1. real   2. increasing

Objective 1

1. radical function   3. polynomial function   5. 64   7. 125

9. -7.389 11. 0.0183

Objective 2

13.

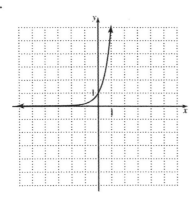

15.

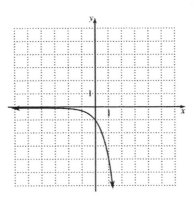

17.

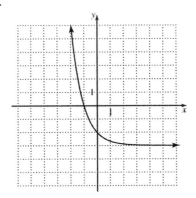

Objective 3

19. -3 21. $\dfrac{7}{2}$

Objective 4

23. The amount after 4 years will be $12,184.03.

9.3 Logarithmic Functions

Key Terms

1. the base b; x 2. exponential; unknown

Objective 1

1. $5^2 = 25$ 3. $3^{-2} = \dfrac{1}{9}$ 5. $\log_{1/10} 1000 = -3$

Objective 2

7. 2   9. 0   11. −4

Objective 3

13. 8   15. 2   17. 25   19. $\dfrac{1}{2}$

Objective 4

21.

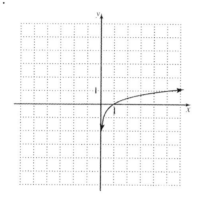

Objective 5

23. The loudness is approximately 7 dB.

9.4 Properties of Logarithms

Key Terms

1. product   2. numerator; denominator

Objective 1

1. $3 + \log_4 11$ 3. $\log_c D + \log_c x$ 5. $\log_3 40$ 7. $\log_x 9x$

Objective 2

9. $\log_4 51 - 3$ 11. $\log_m (m+2) - \log_m (m+3)$ 13. $\log_4 3$

15. $\log_k \dfrac{k}{14}$

Objective 3

17. $10\log_9 s$ 19. $\dfrac{1}{3}\log_6 c$ 21. d 23. $\log_b \dfrac{x^3}{y^{20}}$

25. $\log_c 7 + 7\log_c x - 6\log_c y$

Objective 4

27. The pH is 3.

9.5 Common Logarithms, Natural Logarithms, and Change of Base

Key Terms

1. 10   2. e 3. $\log_e x$

Objective 1

1. 1.7634   3. -0.4034 5. -3 7. $\dfrac{4}{7}$ 9. -3

Objective 2

11. 3.8067   13. -3 15. $\dfrac{3}{10}$ 17. 18

Objective 3

19. 2.1073   21. -0.9269

Objective 4

23. The pH is approximately 3.4.

9.6 Exponential and Logarithmic Equations

Objective 1

1. 3.0650   3. -0.3691 5. 1.0480   7. 1.9597

9. 592.558

Objective 2

11. 2   13. $-38, 38$ 15. 80   17. 4   19. $\dfrac{1}{26}$

Objective 3

21. 4.0 years   23. 1990

Chapter 10 CONIC SECTIONS

10.1 Introduction to Conics; The Parabola

Key Terms

1. standard form   2. left or right

Objective 1

1. vertex: $(-1, -3)$; axis of symmetry: $x = -1$

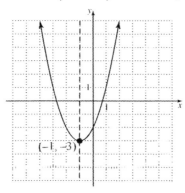

3. vertex: $(-4, 3)$; axis of symmetry: $x = -4$

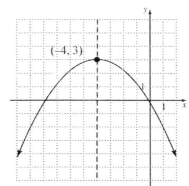

5. vertex: $(3,1)$; axis of symmetry: $y=1$

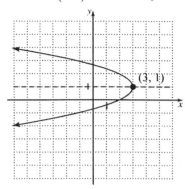

7. vertex: $(-3,-4)$; axis of symmetry: $x=-3$

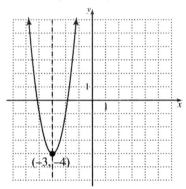

Objective 2

9. $y=(x-4)^2+3$; $(4,3)$

11. $y=2\left(x-\dfrac{9}{4}\right)^2-\dfrac{25}{8}$; $\left(\dfrac{9}{4},-\dfrac{25}{8}\right)$

13. $x=-4(y+3)^2+3$; $(3,-3)$

15. $y=\dfrac{1}{5}(x+3)^2+1$

Objective 3

17. $A=-(w-50)^2+2500$

10.2 The Circle

Key Terms

1. the center   2. distance

Objective 1

1. $\sqrt{85} \approx 9.2$ 3. $5\sqrt{2} \approx 7.1$ 5. $\dfrac{2\sqrt{17}}{5} \approx 1.6$ 7. $(4, -3.5)$

9. $(1.5, -1.5)$ 11. $\left(-\dfrac{1}{28}, \dfrac{25}{36}\right)$

Objective 2

13. center: $(0,0)$; radius: 3

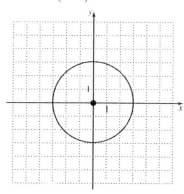

15. center: $(3,-2)$; radius: 3

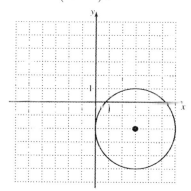

Objective 3 and Objective 4

17. $(x-6)^2 + (y-4)^2 = 25$ 19. $(x+4)^2 + (y-2)^2 = 10$

21. center: $(6,0)$; radius: 6   23. center: $(-2,1)$; radius: $4\sqrt{15}$

Objective 5

25. No

Answers to Worksheets for Classroom or Lab Practice

10.3 The Ellipse and the Hyperbola

Key Terms

1. hyperbola 2. asymptotes

Objective 1

1.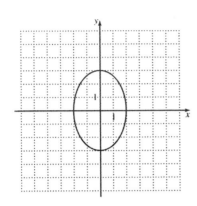

3.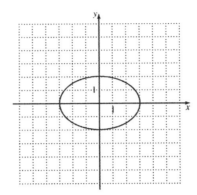

5.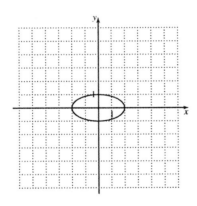

Objective 2

7.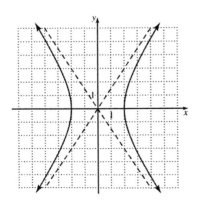

9.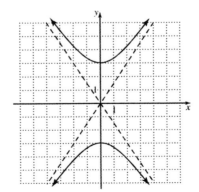

11.

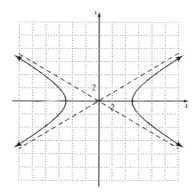

13. parabola   15. $\dfrac{x^2}{64}+\dfrac{y^2}{9}=1$

Objective 3

17. $\dfrac{x^2}{441}+\dfrac{y^2}{289}=1$

10.4 Solving Nonlinear Systems of Equations

Objective 1

1. $(-3,-9),\ (6,18)$ 3. $(3,7),\ (-7,-3)$ 5. $(-3,0),\ \left(2,\sqrt{5}\right),\ \left(2,-\sqrt{5}\right)$

7. $\left(\sqrt{2},\sqrt{14}\right),\ \left(-\sqrt{2},-\sqrt{14}\right),\ \left(\sqrt{2},\ -\sqrt{14}\right),\ \left(-\sqrt{2},\sqrt{14}\right)$

9. $\left(2i\sqrt{3},4\sqrt{2}\right),\ \left(-2i\sqrt{3},4\sqrt{2}\right),\ \left(2i\sqrt{3},-4\sqrt{2}\right),\ \left(-2i\sqrt{3},-4\sqrt{2}\right)$

Objective 2

11. $\left(3i\sqrt{2},4\sqrt{3}\right),\ \left(-3i\sqrt{2},4\sqrt{3}\right),\ \left(-3i\sqrt{2},-4\sqrt{3}\right),\ \left(3i\sqrt{2},-4\sqrt{3}\right)$

13. $\left(10,\sqrt{3}\right),\ \left(10,-\sqrt{3}\right),\ \left(-10,\sqrt{3}\right),\ \left(-10,-\sqrt{3}\right)$

15. $\left(1,4\sqrt{3}\right),\left(1,-4\sqrt{3}\right),\left(-1,4\sqrt{3}\right),\left(-1,-4\sqrt{3}\right)$

Objective 3

17. The length of the smaller bed is 16 ft. The length of the larger bed is 26 ft.

10.5 Solving Nonlinear Inequalities and Nonlinear Systems of Inequalities

Objective 1

1.

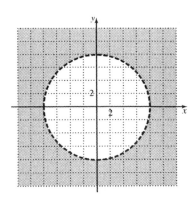

3.

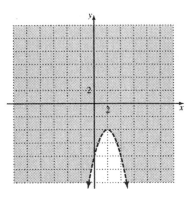

5.

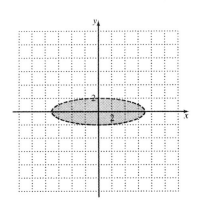

Objective 2

7.

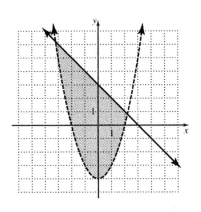

9.

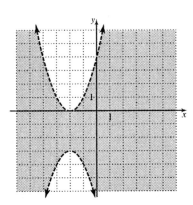